ADI KUNTSMAN AND LIU XIN

# DIGITAL TECHNOLOGIES, SMART CITIES, AND THE ENVIRONMENT

In the Ruins of Broken Promises

BRISTOL
UNIVERSITY
PRESS

First published in Great Britain in 2024 by

Bristol University Press
University of Bristol
1–9 Old Park Hill
Bristol
BS2 8BB
UK
t: +44 (0)117 374 6645
e: bup-info@bristol.ac.uk

Details of international sales and distribution partners are available at
bristoluniversitypress.co.uk

British Library Cataloguing in Publication Data
A catalogue record for this book is available from the British Library

ISBN 978-1-5292-3714-6 hardcover
ISBN 978-1-5292-3715-3 ePub
ISBN 978-1-5292-3716-0 ePdf

Cover design: blu inc
Front cover image: Procreate/Jaz, "Shattered Lives" and
Adi Kuntsman
Bristol University Press uses environmentally responsible
print partners.
Printed and bound in Great Britain by CPI Group (UK) Ltd,
Croydon, CR0 4YY

FSC
www.fsc.org
MIX
Paper | Supporting
responsible forestry
FSC® C013604

# Contents

# List of Figures

# Introduction: How Do We Think About Smart Cities?

At the heart of this book lies a troubling, yet crucial question: how to reconcile the rapid and expansive adoption of various smart technologies that are envisioned as environmentally beneficial, with the extensive environmental harms brought on by digitisation itself? Taking smart cities as one example of such complex relations between digital technologies and the environment, the book offers a paradigm-shifting approach to this conundrum. Instead of accepting the promise of digitisation to improve sustainability, support the environment, and help combat climate change – a promise that underpins many policies, popular representations and much of the academic work on smart cities – the book demonstrates that the promise itself is fundamentally broken. We use the notion of broken promises as a conceptual framework and as a lens to tell a different story about smart cities – and, in doing so, also to pave ways for a new understanding of the narratives, imaginaries, and the material and socio-economic arrangements that configure the relation between technology, human practices, and the environment more broadly.

## The hyper-visible and the invisible environment in smart cities

The idea of a 'smart city' often includes the promise of environmental sustainability, healthy living, and clean energy. These and other environmental benefits are believed to be possible because smart cities are seen as simultaneously green and efficient. Although ecological and climate concerns are rarely the primary focus of smart cities – to the point where many scholars explicitly distinguish between the frameworks of 'smart cities' which focus on urban efficiency and 'eco cities' which prioritise ecological values – many of the world's smart cities' public-facing narratives, such as policy documents or websites, nevertheless tell a story of how smart cities and the digital technologies they draw on, would ultimately make the environment better. They would streamline the collection of waste by using sensor-equipped smart bins which would reduce the driving times of bin collectors; deploy smartphone-operated bike-rental schemes which would replace cars with cycling; monitor air pollution and improve air quality via sophisticated traffic planning; manage city activities and services via publicly shared digital dashboards; and run paper-free e-government services which would reduce deforestation and paper waste. The literature about smart cities, similarly, is heavy in celebratory terms such as 'sharing cities', 'green growth', 'green infrastructure', 'progressive urbanisation', 'sustainable urbanism', 'green technology innovation', 'resilient cities', 'smart future' (McLaren and Agyeman 2015; Galderisi 2018; Kim 2018; Mukherjee 2018; Dastbaz, Naudé, and Manoochehri 2019; Gassmann, Böhm, and Palmié 2019; Tomar and Kaur 2020) and more, where smartness, sustainability, and ecology are often interlinked. And although some smart city initiatives open and complicate the assumptions around what constitutes smartness, in many cases, 'smart' and 'digital' are lumped together, used interchangeably, and celebrated without questions. Cities themselves are often seen as an inevitable development for most of the planet's inhabitants: many academic and policy documents, usually written in highly industrialised contexts which are presented as universal, begin with a statement about how most of the world's population is already, or soon will be, living in a city. In a similar vein, digital technologies are seen as an integral and inevitable part of today's urban living; and are often automatically assumed to be good and equated with social and environmental betterment.

Despite the abundance of specific examples, gadgets and localised 'smart' innovations – from smart bins to road sensors to talking bus stops – what exactly makes the digital environmentally beneficial often remains vague. Are digital technologies good for the environment because using digital tools to run a city is seen as immaterial (see Gabrys 2014) and thus having less footprint on the planet? Are smart cities good for the environment because data affords better understanding of, and adaptation to the changing climate (see Knox 2020)? Equally vague is the sense of what exactly constitutes 'the environment' that would be improved. Is it the 'green spaces', such as parks, trees, green rooftops, and indoor green walls? The quality of air? The consumption of energy? The emissions of carbon? Finally, what is vague is the *where*, *when*, and *for whom* of these perceived environmental benefits. Are they confined to a city district[1] where smart technologies are implemented? What about communities who are not viewed as consumers of smart technologies? What about the areas deemed unsuitable to be 'smart'? What about the global effects of local smart city projects?

To answer these questions – and to move beyond the discussion of whether a particular promise is delivered and to whom – we need to shift our attention from the hyper-visible (the policy promise, the flashy website, the paraded gadget) to the invisible, the ignored, and the epistemologically unimaginable. There are many silences, injustices, and violences which are deeply ingrained yet well hidden in narratives of urbanism and digitisation: the violence of resource extraction; the use of exploitative human labour; the deep inequalities between those benefitting from technologically advanced urban living and those left at the cities' doorsteps and slums; the growing conscription of urban design and digital surveillance in the war on the poor and the racialised; and the global injustices of wealth distribution and vulnerability to the climate crisis. However, unlike the risk of digital surveillance and infringement into urban dwellers' freedom, which does gets acknowledged when smart cities and digitisation and discussed, the environmental harms of smart cities, inflicted on the planet and its human and non-human inhabitants, usually remain unspoken.

This is not due to the lack of evidence or research, though partly can be explained by an absence of a unified terminology or focus. For example, the concept of 'externalities', referring to uncalculated environmental costs, is used in environmental economics but rarely makes its way to other social science research (Fairbrother 2016), and has only recently entered the debates about *digital* environmental harms (Zheng et al 2023). The terms 'footprint' and 'carbon emissions', currently dominating the academic and policy discourse on climate change (Knox 2020; Donaghy 2023),[2] or the research on e-waste and disposal, taking place in health sciences or environmental management, attend only to some types of environmental harms and approach them in isolation. 'Life cycle assessment' (LCA), by contrast, uses a holistic approach to tackle multiple aspects of environmental impacts of digitisation, including devices, infrastructures, energy use and more (Klöpffer and Grahl 2014; Itten et al 2020; Istrate et al 2024), yet is often deployed solely in a technical way, without any consideration of cultural practices, global politics, or social injustices.

When the complexity of environmental harms as *both* technical and socio-political does come to the fore, it happens primarily in fields and contexts that are traditionally not seen as part of smart city research or exist at its fringes, such as discard studies, anti-racist digital justice, or critical work on e-waste or digital labour and their relations to global racial harms and colonial legacies. It is these strands of work that helps us understand why the environmental harms of digitisation remain unspoken in celebratory narratives of smart innovation, or are dismissed with an argument that the benefits outweigh the damages. The harms and damages are dismissed or ignored, we argue, because they take place *elsewhere*, often beyond the spaces of smart cities, and often outside of the Global North itself: the ever-growing extraction of resources needed to produce phones, sensors, and other 'smart' devices; the toxicity and inhumane labour conditions of their production process; and of the e-waste left behind after their short lives, often made disposable by design. Most of these take place in the Global South, destroying lands and communities already devastated by colonisation, post-/neo-colonial extractive economies, and global financial dependence.

How then do we think about today's cities as 'green' and 'smart', from the standpoint of environmental care,[3] as well as from the perspective of global, anti-racist, and environmental justice? How do we approach smart cities' environmental promises while centring our discussion on their environmental harms? In this book, we use the framework of 'broken promises' to address these questions. We use the term broken promises both literally and poetically: broken as a promise that is not kept; and broken as something that is damaged, destroyed, or fundamentally flawed. We look at *what is physically broken in the process of urban digitisation and smartification;* as well as *what is symbolically left out of smart cities' celebratory narratives and imaginaries. What is silenced and deliberately invisibilised? What is promised but never delivered? What is promised in such a way that the promise becomes a trap, a form of violence, where refusing the promise is impossible?* Mapping ways in which questions of digitisation and those of environmental care fail to meet, our book traces the materiality and the environmental costs of smart cities' daily realities, calling the reader to consider: what breaks when a city promises to become 'smart'?

## Broken promises

### *Beyond the gap between promise and delivery*

Recent critical studies on smart cities have noted the gap between the promise of mobility, security, efficiency, and sustainability and their actual implementation (Haarstad 2017).

Ben Green (2019), for example, notes that technologies of smart cities often promise more than what they can deliver; while Alan Wiig (2015) describes that the data-driven vision of urban transformation is an empty rhetoric that works to 'sell' a city in the global economy, rather than actually addressing issues of urban living. The gap between the promise and its implementation, specifically in terms of environmental concerns, was also explored in relation to how the efforts to protect the environment draw on 'smart' tools, whom they engage with in the process, and to what effect. For example, Jennifer Gabrys' (2022) work analyses how smart sensors can be used to contribute to, or hinder citizen engagement with environmental practices. Gabrys describes the 'promissory aspects' of digital technologies whose 'neoliberal sales pitch' has a 'glossy veneer of democratic action' (2022: 15). In reality, however, these technologies fail to have much effect on changing the conditions of environmental pollution or social injustice, while also leading to the proliferation of 'environmentally destructive' digital technologies (Gabrys 2022: 15).

While taking inspiration from these and similar studies, our book goes beyond identifying the gaps between the promises of digital technologies in smart city projects and their deliveries and outcomes.[4] This book is not about evaluating results, presenting 'proofs', or calculating the rates of success and failure. Rather, it is a call to fundamentally reconsider how we approach the relations between digital/smart technologies and the environment, and the hopes and investments that often accompany these relations. To do so, we start with the central paradox of smart urbanism. We argue that promises configure, and depend on, the problems that they claim to solve. Our view of this paradox draws on earlier critical scholarship on digital solutionism (Morozov 2013; Kuntsman and Miyake 2022) that explores cultural, economic, and social conditions in which digital technologies are seen as essential to solving every problem, real or imagined, *including problems, inflicted by digitisation itself.* Our particular focus in this book is on environmental problems, as we inquire how and why digital technologies come to be seen as environmental saviours; how the environment itself becomes intelligible only through digital measures; and how these digital-environmental relations make the harms inflicted by digitisation invisible or even irrelevant.

## Space and time

Our discussion of broken promises foregrounds spatialities and temporalities of smart cities. We historicise smart cities and the actual locales where they are built, to look at which histories are being mobilised to make a promise; which parts of a city are chosen to become 'smart'; which histories are rewritten as the story of urban smartification comes to the fore; which histories are silenced; and which places are left out. In that respect, we incorporate a historical-materialist lens that challenges the timeless and decontextualised smart city narratives and imaginaries. Our attention to space and time helps intervene in, and challenge discursive constructions and architectural visualities of smart cities across the globe, which might appear very similar, but are built on different soils – and have very different relations to their environments.

Spatially, the framework of broken promises situates the narratives and imaginaries of smart urbanism in local contexts. It underscores how smart technologies are embedded in specific geographic, political, and socio-economic environments. Additionally, it makes visible the various demarcations created in the stories and practices of smart cities. These demarcations include the boundaries between the inside and outside of smart cities and smart districts, that is, where smart cities 'begin' and 'end'. And they also include the separation between smart cities and the rest of the world. Throughout the book, we explore how these demarcations are afforded by the assumption that digital technologies are immaterial and environmentally neutral. To challenge these assumptions, we ask about the material consequences of digitisation, and about where and when these consequences become visible.

Temporally, the framework of broken promises allows us to address the multiplicity of temporalities and timescapes of smart cities; the promise that they hold; and the past and future harms that they obliterate. In that respect, our work is both informed by, and challenges, the emerging body of scholarship on temporalities of smart technologies and smart cities. Much of this line of research focuses on the temporalities of real-timeness, and on prediction and pre-emption that are afforded by urban digitisation. Examples include Anne Kaun's and Fredrik Stiernstedt's (2020) work on the temporalities of smart prison, James Merricks White's analysis of how the imaginary of digital solutions to future crisis is used to rationalise smart cities' technological interventions in the present, and Rob Kitchin's (2019, 2023) examination of the multiple temporal rhythms of smart cities, and how they weave together natural, social, and clock times. In these studies, time and temporalities are considered socio-technical phenomena. Foregrounding *environmental* temporalities, we ask what happens to the environment when multiple temporalities of smart cities are at play. Our question pertains not only to what Kitchin described as natural time, but also to the broader question of materialities, environmental impacts, and environmental harms, and how these unfold over time. For example, throughout the book, we look at how the hyper-visible temporality of real-timeness pushes out of sight (and out of mind!) other human-environmental-technological relations, their histories, and futures. We show that the untold histories of resource extraction, or the slow environmental harms of e-waste, water and soil pollution, and biodiversity loss, pave way to a haunted future of digitisation and smartification, a future that exists in the ruins of broken promises.

## *Political economy and affective attachments*

The book addresses broken promises as a simultaneous matter of *political economy and affective attachments*. More specifically, we look at why, despite repeatedly failing, the promise of smart cities has a strong hold. Here, we take inspiration, on one hand, from Karl Marx's (1867) idea of commodity fetishism.[5] Akin to digital solutionism, the idea of commodity fetishism focuses on how the 'need' (or the 'problem') comes to life as part of the commodity creation process, where the commodity serves as the answer to the newly created need. We are particularly informed by critical work on advertising and consumer culture, which reveals how commodities, in the shape of specific discrete objects, are often viewed as having almost magical powers to resolve what is seen as the otherwise unfulfilled need, through possessing that object. Whether it is a particular object deployed in a smart city initiative such as a sensor, a building, or a digitally connected bike, the idea of commodity fetishism allows us to shed light on the political economy of smart city that abstracts the digital from the socio-economic relations and material conditions in which it is produced. Such abstraction naturalises the digital as 'green' and sustainable and obfuscates both the labour involved and the process of fetishisation itself. In the context of neo-liberal governance of which smart city projects are a part, the abstraction often takes the specific shape of isolated experimentations. Experimentations, as we will show in the book, are often devoid of responsibility for what might happen after they are over. They disregard effects that may unfold later, for these would be beyond the experiment itself. Furthermore, they can create relations of impunity, where harms are normalised as an inevitable part of the process, and where failures as attesting to the high-risk-high-return features of innovations.

In thinking about the *affective* hold of failures, we are inspired by Lauren Berlant's notion of cruel optimism: a relation of persistent attachment to an idea that 'is discovered to be impossible, sheer fantasy, or too possible, and toxic' (2010: 94). Berlant's analysis of cruel optimism rests on the idea of objects of desire as a cluster of objects, where object X and object Y are linked through the promise of their desirability. In this book, we zoom in on a range of promises made by smart cities in their planning proposals, descriptions of their vision, and communication documents, to tease out what such objects of desire are, what they promise, and how the promise fails to materialise. We show that the promises of the digital take hold of the collective imagination not only because of the continued politico-economic investment in the IT sector. They are powerful and persistent also because there is a particular hope attached to the promises offered by digitisation. In such promises, a future simply cannot be imagined without the digital – and hopes for environmental betterment are inseparable from expectations of technological innovation. Moving from the affective back to the material and political, we ask, who is most impacted by the cruel optimism of the idea that a digital future is environmentally beneficial?

### Methodologies of 'elsewhere'

Our framework of broken promises simultaneously holds the abstract and the concrete. On the abstract level, we are attentive to the broader political-economic contexts that continue to attach value to fundamentally flawed promises. On the concrete level, we turn out attention to that which is literally broken in the process of making a smart city, or in the aftermath of its failure – be it the depleted land, the discarded e-waste, or the abandoned data. Such sensitivity to the multi-level complexity, in turn, allows us to move past the momentary – the moment in time when a promise is made, or when it appears to be broken – into the longer-term, and often invisible environmental effects of digitisation and smartification. Here we are referring to the 'slow violence' (Nixon 2011) of environmental degradation and to what Nicole Starosielski and Janet Walker (2016) call 'violence out of sight', namely the extractivist nature of seemingly immaterial digital technologies (see also Akese and Little 2018; Aouragh et al 2020), and the harms these technologies inflict on both human and non-human life over time.

Crucially, as we add, these harms are often taking place *elsewhere*, both in another geolocation, and in another time and yet, it is precisely their existence that enables the digital dream of smart urbanism. We are inspired here by Xiaowei Wang's (2020) insightful work on tech in countryside. Wang challenges metronormative thinking in digital research which centres on cities and imagines the rural as being elsewhere, and reminds us that 'rural peripheries, the edges of the world, hidden from view, enable our existence in cities' (2020: 6). To address the *elsewhere*, and the not immediately obvious in smart cities' narratives or ethnographic encounters, we build on and are informed by critical work on racial and environmental harms of digitisation (Qui 2016; Brevini 2020, 2021, 2023a, 2023b; Haritaworn 2020); decolonial perspectives on climate change and rural–urban digitisation (Wang 2020); and critical work in 'discard studies' (Discard studies n.d.) – work that challenges simplistic notions of victimisation and disempowerment, while also foregrounding local and global harms and injustices; work that holds us both inspired and accountable, even when it is not directly related to smart cities, or not directly applied in our analysis.

Being attentive to the harms that take place *elsewhere* – spatially as well as temporally – means paying attention not only to the promise itself, but also to what is left out; what is left behind, such as digital and material ruins of smart cities; or what simply cannot be seen or grasped. In methodological terms, this means that our exploration of two cities, Helsinki and Manchester, took place through an amalgamation of different methods, moving creatively between them. We did not have a pre-existing toolkit, available to us to adopt – or to adapt. Rather, we use this book to reflect on the limitation of existing approaches and methods and on the need to combine insights from many different perspectives; and sometimes, abandon them altogether.

To focus on the social and cultural life of the promise itself, we engaged in close reading of materials collected about the two cities and their 'smart city' projects: websites, maps, policy documents, project reports, media coverage, and interviews. These were mostly the official and semi-official stories told about smart cities by those who were involved with and invested in these projects. At the same time, when considering the invisible and the hidden, we used autoethnographic walking, cycling, and photography in the two cities, as well as online ghost-hunting – looking for traces of what used to be a web- or social media presence. Here we were inspired by Avery Gordon's (1997) work on haunting and the sociological imagination – work that challenges sociology's quest for empirical evidence and turns instead to the social life of the invisible, the unspoken, and the ghostly. Gordon's work is particularly insightful when attending to topics where 'proof' or 'evidence' might be difficult – or epistemologically impossible – to obtain; and where, as a result, intelligibility and legitimacy are attempted through different means – literature, imagination, or haunting. Following Gordon, we reflected on the unseen but 'seethingly present' (Gordon 1997: 8) stories behind our primary and secondary sources: the conditions of digital production such as the extraction or appropriation of resources, and the labour involved, which are rarely present in the stories of digitisation and smart cities; the broken, destroyed, or discarded material objects generated as part of smart city experimentations, which might be hard or impossible to see; and the digital traces of webpages that are no longer available and can be only partially tracked through search engine caches and web archives.

## Our journey to this book

We came to write this book not as experts in urbanism or environmental science, but rather as two interdisciplinary scholars working at the intersection of digital social research, media, and cultural studies, and feminist and anti-racist scholarship. Instead of drawing on – and being constrained by – the canons of digital innovation, urban planning, or sustainable development, we take our inspiration from a broad range of disciplines, fields, and critical voices. The book is a result of several years of collaboration between us, bringing to the table our own research experiences and interests, and the many hours of thinking, creating, and working together. Before writing this book, we explored questions of digital cultures and the environment in China, the UK, Finland, and Sweden, through fieldwork, academic writing, and conversations with students, artists, and practitioners. Individually and together, we are travelling through a dense mix of temporalities and locations, personal experiences and scholarly inspirations.

### *Liu Xin*

My examination of the smart city phenomenon begins with the observation that it is not enough to simply ask what constitutes smartness. This is not only because such an inquiry about smartness often begins and ends with its all-encompassing emptiness, but also because it cannot address the ways in which these promises draw financial investments and shape urban planning and development practices, as well as the environment and the everyday lived realities, in very literal ways. I am more interested in asking what breaks when smart city's promises are broken, how, where, and when.

My interest in researching smart cities began when examining how air quality index became key to the shifting Chinese environmental politics and changing imaginaries of the relation between the environment and human practices. What I found particularly intriguing was how numerical measurements of air quality, produced using different measuring criteria, as well as various sensors and modelling practices, were translated into embodied modes of breathing, sensing and imagining human–atmospheric relations. For example, in examining discussions about air quality index on Chinese social media platform Weibo, I showed that the translation between digital and bodily modes of sensing air materialised the imaginaries of bodies as 'monitors made of flesh' (Liu 2017: 450). Importantly, the process of translation became the site of the emerging Chinese environmental politics. Netizens mobilised air quality index data to voice their concerns and to contest the environmental discourse of the state. They used air quality index to performatively enact, and in so doing, also claim the rights for, citizen political practices such as voting. That, in turn, became the catalyst of the adoption of international air quality standard in China. And yet, the postulation of air quality index as the most accurate and efficient way of knowing air fed into techno-solutionism. It also reproduced an understanding of nature as wilderness that would need to be tamed, and if necessary, waged war on – such understanding, as I discussed in my other work, is also deeply racialised and gendered (Liu 2019).

Another interesting aspect of the air quality index and its data concerns the environmental promises of digital technologies. For example, air quality index apps are hailed for their affordances of near–real–time data. In other words, their promise lies in the transparency and 'live-ness' of data. While the knowledge of the health effects of breathing polluted air is important and useful, it is unclear how exactly the data of air quality solves the problem of air pollution. It seems to me that these apps and the data they produce ultimately target individual behaviour and choices. In the Chinese context, air quality index data is categorised and colour-coded into six levels. The description of each level includes instructions on whether one should, for example, stay indoors, or restrain from outdoor exercises. However, these data say nothing about the measures that need to be taken to mitigate air pollution at the structural and infrastructural level.

In this book, I examine the environmental promise of real-timeness in the context of smart cities. What makes real-time technologies smart and green? Little has been said about *what constitutes the 'real-time' of environmental data, nor what such 'immediacy' does for environmental concerns. In other words, what kinds of environmental problems can be addressed with the near-real-time data? How? What environmental relations are at work in the production, consumption, utilisation and storage of digital technologies and data? What is the relation between the real-timeness of environmental data and the long-term environmental concerns? And to put it simply, what form of human-environment-digital relations is imagined and promised? For whom and what? And what happens if the promises prove to be empty or broken?*

### *Adi*

My interest in smart cities' environmental promises comes from my broader concern with two issues surrounding digital technologies: the rise of compulsory digitality and its role in social injustices, and the impact of digitisation on the environment. In the last decade, I observed the rise of different forms of digitisation, which rapidly became either mandatory or highly desirable. The examples are many: the shift to 'digitally by default' public services, which developed concurrently with algorithmically driven automated decision making processes in areas such as border control or welfare (Kuntsman and Miyake 2022); the normalisation of biometric surveillance in many workplaces (such as the use of fingerprints and facial recognition as a security measure to enter certain workplaces); and the proliferation of smartphone apps that not only govern and manage various daily activities, but increasingly replace in-person services. These, and many other emerging technologies tend to exacerbate existing inequalities such as those around access and disability; furthermore, they are often biased by design, as has been repeatedly shown by scholarship on race and technology (Noble 2018; Williams

2018; Benjamin 2019). Looking at smart cities and their ideas of technological efficiency therefore always takes me back to question of social justice, and issues of exclusion, surveillance, discrimination, and erasure, which are often simultaneously strengthened, and obliterated, by digital innovation. And although smart cities are not a new phenomenon – and, as we note in the next chapter, have roots that far predate recent technological developments – what I am particularly concerned with is how various digital technologies and systems become incorporated into city living in ways that are impossible to escape or avoid.

At the same time, I am acutely aware that while questions of social justice have been raised extensively with regards to urban living, there has been far less discussion of the issues of environmental in/justices of digitisation – and of course, of the relations between social and environmental in/justice. Some of my work has addressed the question of silences around environmental harms of digitisation – or, what I describe elsewhere as a paradigmatic myopia towards environmental unsustainability of the digital. For example, in a systematic review of a decade of work on sustainability and digitalisation that I carried out with a colleague, it was discovered that while digital technologies' benefits for the environment are documented and analysed extensively, their harms are either completely ignored and overlooked, or downplayed as outweighed by the benefits (Kuntsman and Rattle 2019). What particularly interested me were not just the harms themselves – which are widely documented – but the various social, cultural, and political reasons for their invisibility. Increasingly, it became clear to me that such reasons are understood and conceptualised very differently in different disciplines, without any agreed common language. For example, some disciplines focus on digital sustainability as an overarching category, which primarily consists of ways in which digital technologies aid sustainability, rather than tools and methodologies for estimating ways in which digitisation itself is not sustainable.

To remedy that, I have been searching for scholars and practitioners who are addressing carbon footprints of digitisation, in both practical areas such as web design (Greenwood 2021), and in academic research on film and media (Marks 2020; 2023). Tom Greenwood's work, in particular, has led to developing multiple 'carbon footprint' calculators for websites and social media apps offered to businesses and consumers. At the same time, diving deeper into academic work, Laura Marks notes the difficulty to estimate and analyse the carbon footprint of digital technologies, because of many disagreements over what precisely is being measured (Marks 2023). Added to that is what I saw as a very limited understanding of 'footprint' – or rather, *harm* – if focusing exclusively on carbon; and is also detached from principles of social justice. Despite incredible scholarship on discard studies and e-waste, mining and extractivism on the one hand, and on the intersections between racial, social, and environmental injustice on the other, we are yet to see an overarching approach that would speak the language of carbon, extraction, and waste footprint of every digital device, gadget, or platform that shapes our everyday life.[6] And we are yet to see how digital environmental accountability includes questions of global and local racial and social justice.

It was my absolute delight and privilege, therefore, to come across one incredible exception to this general lack: the work by Earth Hacks and partners (Barr-Engel et al 2021; Earth Hacks 2023), who created a report on, and a set of principles for, 'environmental justice in tech'. Outlining a range of harms, specifically on racialised, marginalised, Indigenous communities and those in the Global South, the principles turn to ask how a technology would look if it were environmentally just: freely available and non-profit; anti-racist and anti-colonial; democratic; non-capitalist and non-militarist; intentional and responsible; beneficial to community; and caring for Earth (Earth Hacks 2023). Although focusing on technology more broadly, rather than on digital tech (which is discussed in a section on computer science and software engineering), Earth Hacks' overarching attention to harms resonates deeply with what was driving me in the work on smart cities. *Looking at smart cities' environmental promises for me is therefore a task of noting not only* when *do these promises break, but also* what *they break, what and who must become or remain broken for the smart cities to appear flourishing, and how they might be imagined otherwise.*

## The road ahead

Chapter One, written jointly by both of us, positions our discussion in relation to the literature on smart cities, digitisation, and the environment. We begin by arguing that to decentre and question the ahistorical and apolitical narratives of smartness and innovation, as often presented by IT giants and the tech industry, necessitates a historical account of urbanism, the environment, and relatedly, of sustainability. Approaching the smart city concept through the lens of brokenness allows for making visible the continuities and discontinuities of smart cities with histories of relation between human, social, economic, and cultural practices and their environments. Furthermore, it also contextualises a smart city's promises in specific understandings and configurations of the urban space as a spatial and temporal unit for knowledge production, resource and environmental management, energy generation, and economic development. Our discussion shows how smart cities' environmental promises draw on, differentiate from, and omit specific environmental relations and practices that are discursive, imaginary, and material. To further address the brokenness of these promises, the chapter then turns to the relations between sustainability, digital technologies, and environmental concerns. Here, we outline the many ways in which digital technologies are seen as powerful tools for solving environmental and climate crises, and for promoting a sustainable future. We acknowledge the wealth of literature in this area, and also offer a critique of digital solutionism, where technologies are seen as environmental saviours but their own environmental tolls are rarely acknowledged.

We then move to our two case studies: Helsinki, Finland, and Manchester, UK. Chapter Two, researched and written by Liu Xin, examines the Smart Kalasatama district as part of the 'Helsinki Innovation Districts' project (Fiksu Kaupunki). In this chapter, Liu Xin explores Smart Kalasatama districts online and offline, zooming in on the visual and textual representations of the smart district, as well as past and present energy infrastructures, material histories, and the funding schemes. In teasing out the environmental promises of Smart Kalasatama, Liu Xin argues that the smart district renders the environment experimental, fragmented with internally dehiscent and hierarchically positioned temporalities and spatialities. Chapter Three, written and researched by Adi, focuses on Manchester, its two recently completed smart city projects, and its latest smart city strategy. Looking at a number of promises, made throughout the projects, the chapter traces visions of Manchester as a smart city as they take shape between the municipal, the academic and the corporate, involving a range of actors and narratives. One of the key themes that runs throughout the chapter is the disconnect between the digital and the environmental agendas; another is the lack of continuity and legacy between the different initiatives. Bringing the two together, the chapter ends with a proposal to look at ruptures, remains, and ruins, to better understand the hopes and the violence embedded in smart city promises in the city.

Our thoughts and analysis come together in the jointly written Conclusion, where we return to re-evaluate the notion of broken promises as a conceptual tool to assess the complex relations between smart cities and the environment. In thinking about smart cities as both material and imaginary entities, we ask, what is hidden from view or shut out of smart cities' dreams and visions? What is physically or symbolically discarded, ignored, disconnected, or destroyed? We also ask, what is left, when smart cities' promises do not materialise, and what ruins lay the ground of smart cities' broken promises? Our concluding chapter begins with the notion of the cruelty of failure, when failure is normalised through optimism and is built into a promise that was never meant to be delivered. Bringing our two case studies together, we then look more closely at broken geographies and broken temporalities of smart cities. After discussing, what *breaks* when cities attempt to become smart, we turn to looking at what *remains*, what is left after – after a smart city project is finished, after an experiment is over, after a technology becomes obsolete. We turn to processes of digital and material ruination, to discuss the aftermath of smart cities, as a perspective that moves us out of the short-termism of techno-innovation hype, into thinking about future accountability, care, and social and environmental justice.

# ONE

# Smart Cities, Digitisation, and the Environment

*Imagine the following building: it is multi-functional, economically profitable, environmentally efficient, and marketable. It provides spaces for apartments, offices, parking, and exercise. It optimises the relation between the natural and built environment by utilising natural light and air circulation. It is self-sufficient, generating its own electricity and heating from within the building. Its design allows for safety, convenience, and easy mobility, being remotely monitored and controlled. Its unusual and striking appearance and its advanced technologies and operating systems increase its public visibility, and generate revenue for its owners, architects, and construction and service companies that support it. The building becomes an object of marketing, a tourist attraction, and a testimony to the success and trust-worthiness of its stakeholders.*

*Such a building would fit the description of many smart city projects that exist around the world today that claim to respond to two emerging concerns of urban living. First, urban centres are expected to further expand to accommodate an additional 2.5 billion people by 2050; and second, urban spaces face challenges brought by the multiple, layered, and often simultaneously unfolding crises of climate change, environmental degradation, pandemic, and geo-political conflicts. And yet, it might be surprising that the imaginary building just presented is based on the description of the Woolworth Building – the monumental skyscraper in New York City that was completed over a century ago, in 1913; and was the world's tallest building until 1930. The building 'could generate up to 1,400 KW, lighting thousands of* bulbs, running ventilating fans and 850 exactly synchronised clocks, and rushing twenty-seven high speed elevators up and down the sixty storeys. The exhaust steam from electricity generation was used to heat the building' (Calder 2022: 310–311). Commissioned by Frank Woolworth, the founder of the Frank Winfield Woolworth company, the building was designed to be architecturally striking, extravagantly luxurious, and profitable. It was set to be a company headquarter that would demonstrate its success and attract tenants who could afford high rents and was called the 'Cathedral of Commerce'. The building was heavily marketised, with a spending of $100,000 for its campaign and coverage by more than 200 newspapers worldwide already more than two years before its opening. The opening ceremony of the building showed off and added to its spectacular appearance. With a remote switch, the then new president of the United States Woodrow Wilson turned on the 80,000 electric lights on the building's exterior.

The lighting of the Woolworth Building embodies the rapid transformation in electrification and energy infrastructures during the late 19th and early 20th century in Europe and North America – a history that is also marked by colonial and imperial exploitation and violence. In the shadow of the hypervisible and enchanting illuminous electric landscape of New York lies the deeply unequal access to heat and electricity both locally and globally, mining in colonial and settler territories to meet the intensified energy demands, increased air and water pollution, as well as the racialised exploitation of labour (see Arnold 2005; Santiago 2013; Hasenörl 2018).

Woolworth Building's sophisticated and innovative design can be seen as the predecessor of today's smart city architecture. It is profit-driven and is preconditioned by the invisible violence of colonial extractivism and exploitation. At the same time, it is envisioned as an innovation in self-sufficiency and sustainability, although without necessarily elaborating who and what such sustainability benefits. Like the Woolworth Building, smart city buildings of today are seen as utilising and maximising every resource possible, whether it comes from within or outside of the building; whether it is 'natural', such as air or light, or technologically manufactured, such as electricity or heat; whether it is justly acquired; and who or what it might have harmed *in the process. And like the Woolworth Building, smart cities of today obscure the harms they rely on, both human and environmental.*

## Introduction: decentring the digital, centring the environment

We situate our discussion of broken promises of smart cities by taking a closer look at academic and popular imaginaries of smart cities and their relations to the environment: how exactly are these relations understood, what kind of environment is envisioned in smart city's practices, and how such environmental promises are said to be delivered. Throughout the book, we approach the environment as simultaneously the actual materiality of both 'natural' and build environment, *and* the way these are understood, perceived, and lived in various social, technological, economic, and political contexts. As the case studies in Chapters Two and Three will show, while smart city projects frequently underscore the environmental benefits of digitalisation, what exactly constitutes the environment is often left vague. In our analysis of smart city narratives, visions, and realities, we will focus on both the material and the discursive, allowing us to trace the rise (and fall) of various visual representations and narrative logic, as well as to pay attention to what is taken for granted, assumed, or concealed, and when. Take for example the image of 'the blue planet' from 1968: the Earth as seen from Apollo 8. The photo by NASA (Moran 2018) showing a view of the Earth from space was significant in consolidating the imaginary of the environment as one Earth. It shifted the concept of the environment from the post-war concerns of management and regulation of resources bounded in national spaces to a planetary scale (see for example Warde, Libby, and Sörlin 2018). The image and the term 'blue planet' have since become iconic.

Smart cities, similarly, too, have their established visual representation. For example, if we look for smart city websites, webpages about smart cities on various tech companies' websites, or even input 'smart city' into Google image search or into a stock image database, we will find striking iconographic similarities. The images that come up show a generalised, sky-scraper-ridden urban landscape, placed on a blue or blue-green background, often depicting a light-filled night sky. Above the buildings we see floating icons, some are immediately recognisable within the smartphone and computing interface lexicon – a sign for cloud storage, a sign for Wi-Fi – and others less familiar (see for example Acea Waidy Wow n.d.; TWI n.d.). Some icons refer to specific 'smart' arrangements within a city: a smart car, a smart light, a smart bin, a sign for a power station or a smart water management system (see for example CB Insights 2020). Many of the images also depict connectivity as flows of data as lines between the buildings or the icons themselves. The connectivity sometimes extends beyond a cityscape into a global scale, by showing airplanes, and occasionally also the 'blue planet' – this time, however, the globe is used to visualise global connectivity rather than planetary environmental care.

In this iconography, the digital, to paraphrase Donna Haraway (1998), is portrayed as being nowhere and everywhere at the same time. This makes the flows of data and connectivity hypervisible, while simultaneously obliviating the environment (unless it appears as a resource that is managed in a 'smart' way). The materialities and, in particular, the material harms of smart cities, such as resource extraction or the carbon footprint of urban digitisation and smartification, usually go visually unrepresented. Furthermore, they are often erased from smart city visions, as we demonstrate throughout the book. Invisible in the websites and reports of smart cities, filled with glossy images and sleek infographics, they are not shown in popular narratives of technologically-enhanced urban living. Equally, they are absent from the imagination of everyday citizens and decision makers alike. There are many political, economic, and cultural reasons for that invisibility which we unpack towards the end of this chapter and throughout the book; and it is worth noting also that these materialities are complex to begin with. For instance, the material implications of a mobile app concern not simply how the mobile phones themselves are manufactured, or the e-waste they generate at the disposal stage, but also, what goes into the process of producing and storing their data, what forms of energy are used and what are their sources. What is crucial in being attentive to the harms and their complexity, we argue, is to shift our focus: we must not simply make the material and environmental implications visible, but bring them into the centre of academic and public conversations about smart cities.

The task of this chapter, therefore, is to change the focus from the digital to the environment, as a starting point in discussing smart cities and their promises. Instead of accepting and reproducing the story of smart city as the story of digitisation and innovation, this chapter re-reads smart city narratives by centring the environment, and in so doing paving the way for decentring the digital as a taken-for-granted environmental saviour. We do this in the following two ways. First, we ask how the environmental conditions are figured in, and shape the established narratives of smart city's digitisation practices, as well as what and who is privileged or excluded from these narratives and practices. In other words, we discuss the environment not as a given field where digital technologies operate – instead, we trace how the environment might provide possibilities and constraints for digitisation. Second, with the starting point that smart cities are projects that involve changes in built and natural environments, often beyond what is considered 'digital', we question whether the digital is the defining parameter of the environment, as often implied in policies or initiatives, such as the 'green digital transition' and the language of 'green and digital future' (Muench et al 2022). We are questioning the use of the digital as the logic, framework, vocabulary, and measurement through which the environment comes to be known and made sense of: for example, that without data one cannot comprehend the size of green spaces or feel the level of air pollution.[1] Instead, we tease out the contingencies of the relation between the digital and the environment that are shaped by and embedded in various economic, social, and political relations as well as in tangible, material realities.

As we will show in the case studies in Chapters Two and Three, these contingencies – as forms of broken promises – are manifested in material and temporal incompatibilities between the seamless immediacy promised by digital processes, and the long-term material consequences of urban development that often unfold slowly. We argue that behind the idea of a universal model of 'smart city' lie histories of technological innovations as well as their failures, abandoned initiatives, and decaying ruins; visions and technologies of energy transition and related infrastructural changes and their multiple, long-lasting, and often unpredictable consequences; green and brown spaces, on which digital infrastructures are built or are about to be built, and people and communities that occupy them – or are/might be displaced from them. The contradiction is further seen in the discrepancies between the logic of environmental sustainability and the broken temporalities of project economies that rely on limited cycles of funding schemes, and often come to a halt or disappear completely when the project/funding ends. Such project economies, as we will show in the following chapters, often operate simultaneously on multiple local, national, and regional levels. These levels at times intersect, and at times also clash. And finally, we show that the classed, racialised, and colonial histories of urban developments and digitisation are often elided by, but nevertheless haunt, smart city practices.

Such decentring of the digital that we propose here also compels us to ask where the digital begins and ends, and what happens if the story of the smart city is told not from the vantage point of international corporations and IT giants, consultants, or governments, but from the matter of environmental preconditions and implications – such as ecosystems, atmosphere, trees, minerals, species – which are located within and beyond the measurable or identifiable boundaries of smart cities, and shaped by social and political relations. To do this, the chapter proceeds as follows: first, we historicise and contextualise the 'smart city', making visible how the changing configuration between human and environmental relations have taken shape through specific historical and material conditions, rather than simply emerging as the effect of smart city practices and the digital technologies that operate in them. Second, we examine how the environment is understood in smart city narratives, teasing out the difference and relation between green and sustainability. Finally, we move from smart cities to the question of sustainability more broadly, to examine and critique ways in which digital technologies are imagined as the key driver of environmental sustainability, while their own unsustainable nature is persistently ignored.

## Historicising and contextualising smart cities

The story of smart city revolves around two themes: digitisation and urban development. It goes like this: smart cities emerge as a result of ongoing innovation and development of digital technologies; which, in turn, are supported by, and grow in, smart cities. New technologies generate and analyse data from urban spaces, which in turn become the testing bed of these technologies. Although potentially resemblant of the Woolworth Building and many other visions of modern life transformed by innovation, 'smart city' as an idea is distinctly tied to the tech industry. While the idea of smart urbanism emerged in 1990s, IBM's registration of the trademark 'smarter city' on 4 November 2011 is often seen to mark a significant starting point, if not the origin, of smart city development. In other words, the story of smart city is primarily a corporate story. It features the vision of IBM as a global leader of IT, evidenced in the title of the talk by IBM's former CEO Sam Palmisano in 2008, that kickstarted a smart city technology market – 'A Smarter Planet: The Next Leadership Agenda' (Palmisano 2008). IBM defines a smart city as 'one that makes optimal use of all the interconnected information available today to better understand and control its operations and optimise the use of limited resources' (TWI n.d.).

Although this may not be the working definition of all smart city initiatives and projects, it is telling of two key issues of smart city imaginaries that are central to the discussions in this book. First, as a corporate strategy for increasing competitiveness and creating a niche market within the tech industry, the smart city narrative identifies the problems of urban development and their solutions as essentially informatic and digital ones. Second, and relatedly, a smart city's narratives also frame environmental questions in specific ways, which are tied to the technological solutions offered. In the case of IBM's definition of smart cities, for example, environmental questions are defined in terms of limited resources – such as energy and water – that are key for urban infrastructures. In this logic, the environmental promise of smart cities lies in the optimisation affordances of information and digital technologies that are seen as the best – and often the only – way to address the problem of shortage of resources. These promises prioritise market-driven efficiency over nature and ecology, *even when* they speak the language of environmental care. They also flatten the differences between, and specificities of, the smart cities themselves, in terms of their social fabric and their actual environmental and material conditions. What, then, happens to the environment as it is subsumed into the promise of technological innovation? What might these promises break or leave behind?

Using broken promises as an analytical lens allows us to think through these questions and to approach the configurations of smartness and environment not as universal but as politically, economically, historically, and contextually specific. Looking at the corporate origins of smart city narratives in such way allows for posing questions such as whose interest is represented in these narratives, what assumptions are made about life in smart cities, and what is elided in each specific story or image.

Such an exploration allows to focus on what conditions a specific narrative, what kinds of continuities and changes are discernible, and what alternatives are possible. This approach differs from the existing research on smart cities that usually begins and ends with the equation between smart cities and digitisation of urban spaces. In focusing on what is promised and what breaks, we are informed and inspired by critical scholarship on urban digitisation that examines how various forms of digital surveillance and control are normalised as part of urban 'smartification'; and how existing injustices and exclusions are exacerbated by digital technologies (Williams 2018; Sengupta and Sengupta 2022a, 2022b). This scholarship shows how smart cities produce what Morgan Mouton and Ryan Burns (2021) describe as digital neo-colonialism. They argue that open data, seen as a cornerstone of a smart city's digitisation practices, 'serves as a key mechanism for digital neo-colonialism, in that relations of dominance and extraction are justified through elements of platforms such as database structures, data models, relational schema, and acceptable nomenclature' (Mouton and Burns 2021: 7). Digital colonialism, here, is simultaneously about legitimising the reign of corporate digital and data-driven solutions, and about excluding and obfuscating other approaches to the city.

This important critique shows that it is imperative to understand, plan, inhabit, and transform urban spaces beyond the logic and practices of data and digital solutionism. However, the alternatives proposed still revolve around the digital: they put forward ideas of digital sovereignty, digital democracy, or data justice, yet fail to challenge the underlying assumptions and imaginaries of the digital as the environmental saviour in the corporate storytelling about a smart city. What is thus left unquestioned is how the digital comes to be seen as not just the only option, but a magical one – its decisive role in shaping urban futures lies in its capacity to connect and optimise everything and everyone, even as, and precisely because of, its perceived *immaterial* nature. In other words, it is seen to be ephemeral and yet with significant material implications, a detachable add-on isolated from nature and humans, but one that can somehow fundamentally reconfigure various human-environment relations into one efficient and integrated system.

This is not to say that the digital is not one of the most important elements of smart city visions. What we argue though is that alternative imaginaries of smart cities necessitate decentring the digital beyond a narrow understanding of a set of technologies that can be simply introduced and teleported from place to place. For example, instead of simply asking about the consequence of smart city's digital practices, it is crucial to examine the historical, material, and contextual specifies of specific smart cities, as well as the various practices and infrastructure changes that do not include what is typically considered digital. As Håvard Haarstad shows, smart city projects necessarily include many non-digital, 'commonplace' practices, such as 'retrofitting buildings, introducing electric buses, or creating distributed forms of energy production' (2017: 431). These are rarely considered innovative until they are supplemented with *digital* technologies that generate data about and from these practices. And here lies one of the most crucial mechanisms of the circular logic of digitisation and its gripping power: the digital makes non-digital/non-datafied solutions invisible, maintaining new technologies' forerunning status where innovation (or efficiency) is collapsed into the digital, because other practices are simply not seen or recognised as existing in, or relevant to the digitally dominated world.

Another crucial obfuscation in discussions of urban digitisation – and indeed, one of the most violent impacts of digitisation – is the question of what digitisation consumes. For example, the development of lithium-ion batteries that are key to the electrification practices of wireless technologies, smartphones, and electric grids, all of which are imperative for a smart city's actual existence, necessitates a massive scaling up of extraction, processing, and production of raw materials, metals, and chemicals. These processes are in themselves energy and/or land and/or water intensive and often result in large amounts of waste, many of which toxic, that have long-term implications for biodiversity and the livelihood of workers and local communities. The production of lithium in the Salar de Atacama – Chile's largest salt flat – is a good example that illustrates this point. Developed by the Sociedad Química y Minera de Chile (SQM), the Salar de Atacama provides nearly 40 per cent of global demand for lithium in 2002 (Méndez 2005). The lithium production at the Salar de Atacama is hailed to be 'sustainable', as 90 per cent of its brine-based extraction is said to be from solar energy, which is less energy intensive than traditional hardrock mining. However, as James Morton Turner (2022) shows, the brine-based lithium extraction is water intensive, and has immense social and environmental consequences. The privitisation of land and water, and the favourable legislative and economic conditions for mining, disrupt the farming practices and communal irrigation infrastructures of the indigenous Atacameño communities (Blair et al 2023). By 2010, the volume of the groundwater used for the lithium production at the Salar de Atacama is equivalent to approximately one-fortieth of annual water usage of New York City. The groundwater is vital for the survival of the tamarugo tree, the only tree that survives the climate of the dry salt flats. 'Researchers warned that groundwater pumping would create a cascade of ecological effects, jeopardising the tamarugo trees, the biological productivity of the salar's few lagoons, and the habitat of well-known species such as the flamingos' (Turner 2022: 356).

It is important to note that these environmental implications are also social and political, embedded in historical power asymmetries that are racialised, classed, and gendered. For example, the lithium production in the Salar de Atacama utilises the mining infrastructures with a long colonial and imperial history, extending from the Spanish colonial era into the post-WWII period of American 'stockpiling [of] strategic resources like uranium and lithium for the development of nuclear energy and weapons, as policies of resource nationalism became foundational to the consolidation of global power' (Blair et al 2023: 5–6). Such power asymmetries take place not only in the extraction process but also during the process of battery production, which often takes place at a different site than the resource extraction – in this case, in China. To minimise the costs, the Chinese company BYD, one of the biggest global lithium-ion battery suppliers, substituted machines with low-cost human labour working long-hours, seated along the assembly line in dry rooms. Most of the workers are young women from Chinese rural areas.

As the outlined examples show, the human and environmental tolls of smart cities are *the* material condition for digitisation and electrification, and yet are entirely absent from both the visual representations and the narratives of a smart city's environmental promises. Smart urbanism's approach to the environment, in that respect, is similar to that of its predecessor – the industrial city of the first industrial revolution, embodied by the Woolworth Building. Both obfuscate the violence of resource extraction, labour conditions, and colonial theft. Both follow the modernist urbanist understanding of nature as in need of human control. The two differ however in how the control is imagined and materialised. An industrial city controls nature by subordinating it and turning it into 'resources' and 'spaces' for industrial and urban expansion. A smart city, similarly, controls and manages nature, employing it as a (natural) infrastructure and testing ground of data. Furthermore, smart cities use nature *through the lens and power of the digital.* In smart cities, the digital acts as the defining parameter that renders some environments and human-environment relations measurable, perceivable, and therefore, valuable, while invisibilising and devaluing others that cannot be captured or datafied. Understanding smart cities' relations to nature in this way gives us a new angle to examine the promise of sustainability. In the next two sections, we take a closer look at the ways in which smart cities disavow their historical and material specificities through universalised visions of digitisation and how the digital itself persistently appears as green and sustainable, obscuring its own environmental harms.

## Green and/or sustainable cities?

Sustainability is often seen as a key feature and result of smart city projects. However, in the descriptions of smart city projects, sustainability is used in self-evident, self-referential, and self-explanatory ways. *Smart city is smart because it is sustainable.* In effect, both smartness and sustainability are what Mark Davidson (2010) calls 'open signifiers' that are mobilised to encompass a wide range of ideas and practices. According to Davidson, sustainability emerged as a catch-all ideal in urban and social policy in the early 2000s. In a somewhat similar fashion to the ascendence of the concept of diversity that replaced notions such as equality and justice, seen as overly political (see Ahmed 2012), sustainability is used in response to the emerging social, economic, and environmental concerns in an era of a shrinking welfare state and the rise of neoliberal values. The vagueness or lack of precise utility of the concept of sustainability allows it to function 'as a foil to give the appearance of doing something about global warming and the environment, when in effect it is largely deployed to maintain the priority of economic growth for achievement of global competitiveness' (Gunder and Hillier 2009: 20).

While the analysis of sustainability as an empty signifier is important, we suggest that it is not enough to simply diagnose the vagueness or ask how it is framed in various smart city discourses. Instead of simply asking what sustainability is, we suggest that it is imperative to trace what it *does*, precisely when it is vague; when its details are absent; or when the term itself is overused. Doing that necessitates examining the various tensions and fissures between a smart city's environmental promises and its other priorities, and in particular, the incompatibility between ideas of urban environmental sustainability, especially those focusing on resource efficiency, and ecological care for land and biodiversity. For example, as Federico Cugurullo (2021) shows in the case of Hong Kong's smart city projects that involve both the Central Policy Unit and the Environment Bureau, the latter's ecological expertise and concern for biodiversity was silenced. The construction of 'smart' and 'green' buildings that promised to optimise energy and resources use did not take broader environmental consequences into account. As Cugurullo explains:

> the building process was not informed by any ecological study. The developers did not conduct an environmental impact assessment, and simply cleared the plot of the existing vegetation. They then levelled and paved the land, ultimately creating a large expanse of concrete and steel, insensitive to issues of biodiversity loss and ecosystem services. (2021: 108)

Crucially for our discussion in this book, Cugurullo's observation identifies important discrepancies between different understandings of environmental sustainability itself: it can be about economic profit when it focuses on optimisation of resources and energy, or about broader ecological concerns such as biodiversity and ecosystems. We draw on, and further develop such analysis of sustainability discrepancies by asking not simply what environmental impacts smart city projects have failed to acknowledge, or intentionally silenced, but more specifically, how, and why, the changing configuration of the environment in the smart city's imaginaries allows for such an erasure precisely *as* the environmental promises of the digital. To do so, we turn to the two key terms typically deployed in the smart city's descriptions of the environment – green and sustainable.

As we will show in the following two chapters, the two terms are often used interchangeably. However, the references to what constitutes the environment and its function differ from case to case, shaped by changes in economic conditions and political climates. For example, the use of sustainability in smart city discourse has been conditioned upon, and in many cases demanded by, various policy initiatives and funding schemes, which dictate a focus on sustainability as part of their overall strategy, often framed in the language of SDG (sustainable development goals) as used by the UN (United Nations Department of Economic and Social Affairs 2022). This only sometimes includes an explicit focus on the environment and climate change, but, more often, uses all forms of sustainability listed in the UN 17 SDGs, whereby sustainable is assumed to be green, and the latter assumed to be caring for the environment. To contextualise and unpack these conflated assumptions, we need to historicise the use of these terms, starting with the use of 'green' within the context of urban planning. This will help us to better understand how they inform today's smart city's environmental promises.

The emphasis on 'green spaces' in planning industrial cities was most pronounced in the Swiss architect Le Corbusier's work in the early 1900s. The model of 'Corbusier's ideal city (1887–1965)' is one of the most influential types of modern urbanism that has profound implications for how nature and environment are understood in contemporary urban planning. Critiquing the model of industrial cities beginning in the industrial revolution in the late 1800s, Corbusier argues that green spaces must be planned and built into urban spaces. As he writes in *The City of To-morrow and Its Planning*, first published in 1929:

> Here is a clear aim before us: the city of Tomorrow could be set entirely in the midst of green open spaces. The mistake made in New York was that the skyscrapers were not built in the parks … We must increase the area of green and open space; this is the only way to ensure the necessary degree of health and peace to enable

men to meet the anxieties of work occasioned by the new speed at which business is carried on. (Corbusier 1929: 145–177)

The main purpose of inserting green, with '*proportional mean*' in urban spaces is however not so much for the sake of environmental protection or care for nature, but to 'satisfy man's accustomed needs, and bring him joy, recreation, beauty and health' (Corbusier 1929: 141, emphasis in original). In Corbusier's model, the urban space consists of geometrically ordered built environment and abundant but controlled green areas – 'a city of towers placed in a vast park' (Cugurullo 2021: 29). In other words, the green, in his vision, is not natural, but built, and controlled. At work in Corbusier's model of urban planning, and similarly to the industrial city he criticises, is 'a conceptual fracture (and limitation) of modernism in which the *natural* and the *urban* are understood as oppositional and ultimately incompatible types of space which cannot exist in the same area' (Cugurullo 2021: 29, emphasis in original). It follows then that, despite its emphasis on green spaces, the Corbusier city is not unlike the city of industrial revolution in the sense that it, too, understands the natural environment as needing to be controlled, shaped, and used by the humans. Whereas the industrial city sees nature as fuel and raw material, the Corbusian vision of a modern city confines nature within a designated green area, a controlled resource to satisfy the city's inhabitants.

The concept of sustainability did not appear in Corbusier's writing. The concern for ecological limits and their transgression that informs the contemporary understanding of sustainability began to take shape in the late 19th century. In the field of urban planning, its roots could be traced to Ebenezer Howard's (1850–1928) formulation of garden city which proposed the conservation of nature (Basiago 1996; Howard 1898). While a couple of other models of urban planning that echo the garden city were developed, it was not until the late 1960s and early 1970s that the linkage between ecology and sustainability became explicitly established in eco-urbanism (see for example Basiago 1996). Although eco-urbanism and smart urbanism are often considered as two different ideals of urban planning, with emphasis on the environment and the digital respectively, they increasingly intersect. Ecological urbanism emerged as a countercultural movement against urban development. It draws on the science of ecology which puts emphasis on the importance of respecting the environmental limits of urban development, as well as the impacts of built environment on wider ecosystems. However, it, too, privileges the design and built environment in service of economic interests such as 'new, decarbonised iterations of capitalism as the only hope of our collective urban future' (Caprotti 2014: 1287).

Whereas green spaces in the Corbusier model are seen as trees and parks that can be proportionately inserted in urban spaces to increase residents' well-being, in ecological urbanism they are hierarchised according to their capacity for sustainability. As Cugurullo writes, plants 'that can be used as sources of food and energy, as well as for pharmaceutical purposes' (2021: 34) are prioritised, whereas those considered ornamental and unnecessary are taken away to create the space and energy needed for developing agriculture in urban spaces.

Here, it is important to note that the acontextual and ahistorical description of sustainability in contemporary discourse – 'the idea that to endure, a society must not undermine the ecological underpinnings on which it is dependent' (Warde 2018: 4) – misses the crucial point that sustainability is an economic concept. Central to sustainability are two interdependent measurements – carrying capacity and sustainable yield. The notion of sustainable development came from the mechanical theory of '"maximum sustainable yield" conceived by (ecological) fishery science in the 1950s, which itself came from the notion of "sustainable (nachhaltig) management" developed by German forestry science in the eighteenth century' (Bonneuil and Fressoz 2016: 53). Furthermore, the temporality of sustainability is not simply long term (which has also increasingly become an empty signifier mobilised in various climate delay strategies), but one defined by a relation of optimisation that produces a balance between two variables – certain natural resources and certain human needs. The mechanical theory of sustainability revolved around optimising natural resources conceived as 'linear and reversible' (Bonneuil and Fressoz 2016: 115). As Christophe Bonneuil and Jean-Baptiste Fressoz write, the ultimate concern of sustainability is to perpetuate economic growth. In this context, 'the environment became a new column in the bookkeeping of big corporations, which gave themselves new sustainable development divisions' (Bonneuil and Fressoz 2016: 53–54).

Interestingly, while green and sustainability are elaborated as goals to be achieved in Corbusier's utopia city and ecological urbanism respectively, they are assumed as expected outcomes in smart city urbanism. For example, as Håvard Haarstad observes, environmental sustainability does not play a lead role in the discourse on smart cities at the EU-level, but 'is largely an assumed result of more efficient, cost-effective urban systems and greater availability of data. It is not jeopardised by economic growth but is dependent upon it' (Haarstad 2017: 429). In the following chapters, we closely examine whether and how the ideas regarding 'green' and 'sustainable' are depicted in, and inform smart city visions in Helsinki and Manchester. Before proceeding, however, we must first dwell on the relations between sustainability, environment and the digital, to better understand both why digital and smart technologies are adopted to support green and sustainable goals of smart cities, *and* why their own environmental harms are so often ignored in smart city visions.

## Environmental sustainability and the digital

If 'green' and 'sustainable' are often interchangeable, 'digital' and 'sustainable' frequently appear together, in a relationship that Sara Ahmed has describes as 'sticking together [of] signs' (Ahmed 2004), where words, objects, or emotions regularly appear together, and often develop a relationship of metonymic substitution. Here is one telling example. In the late 2010s, *Sustainability Science* journal published a special issue, titled 'The game-changing potential of digitalisation for sustainability: possibilities, perils, and pathways', where digital technologies 'in the form of e-health services, robotics, or emission reduction solutions' were celebrated for their anticipated ability to 'help individuals, organisations, and nations achieve a more sustainable planet in light of the sustainable development goals' (Lock and Seele 2017: 183). A brief web search for popular science publications, or tech companies' PR, quickly reveals a similar sentiment, whether it is the idea of 'digital transformation' (Hughes 2020) or 'data and digitalisation'as 'key tools'to tackle climate change (Álvaro-Alonso 2021); or that 'connectivity and sensor technology help create a greener world' (Thales 2021). While much of such corporate communication can be seen primarily as a form of greenwashing – a rising phenomenon in the tech industry which is finally recognised as a serious environmental culprit – the sentiment regarding the endless possibilities of digital technologies *for sustainability* and the environment is much deeper, and much more endemic.

To understand the 'stickiness' of digital and sustainable in the context of smart cities, it is not enough to see how smart cities consistently equate smartness with sustainability and simultaneously also with the digital (cities become smart as they become digital; and becoming smart/digital makes them sustainable). To understand the brokenness of smart cities' environmental promises, we also need to unpack and interrogate why sustainability is so often understood as being *dependent on* the digital, such as the Internet of Things, Big Data, AI, or digitisation more generally – especially despite a wealth of evidence about environmental tolls of digital technologies and ICT (Sengupta and Sengupta 2022b). Several years ago, one of us carried out, with another colleague, a systematic review of academic articles on environmental sustainability and the digital, published between 2008–2018 (Kuntsman and Rattle 2019). We found that in the majority of publications digitisation was seen as being in service of sustainability projects – from technologies of sustainable innovation and eco-efficiency to powerful tools of gathering and communicating information about the environment, including environmental changes and harms as well as 'greener alternatives'. Our findings were not limited to the corpus of articles we had reviewed: the sticking together of sustainability and the digital continues to appear in various sustainability projects and research to this day. What's more, a very similar logic which was observed in the systematic review also operates in many projects and narratives of smart cities, as we show in the next two chapters. Let us have a closer look at the logic behind celebratory adoption of digitisation, by unpacking what frameworks render the digital key to sustainability.

The first framework, frequently seen in sustainability studies and policy (Kuntsman and Rattle 2019), and endemic to the fields of environmental communication and environmental education more broadly, approaches digital technologies as tools for changing people's minds and behaviour. The 'change' here is understood very broadly, from awareness and education, which in turn, are believed to lead to shifts in individual behaviours which would help the environment; to wider social changes, including collective mobilisation and activism. For example, scholarship on education for sustainability celebrates digital technologies such as apps, mobile devices, and on-line classrooms, as effective, engaging, and thus transformative tools that can be used to teach children and young people about the environment and climate change through a variety of interactive practices, 'location-based' and gamified learning (see for example Howard 2015; Schaal and Lude 2015). The promise of behavioural change, embedded in digital educational tools, is ever more prominent if we turn to the field of sustainable consumption, which rests on the assumption that consumer culture and consumer practices can be environmentally damaging, but equally can be changed to be more sustainable – this is, of course, with the help of digital tools such as smartphone apps that calculate one's footprint, inform us about sustainable products and bring awareness of one's consumer behaviour. Offering readily-available information on more sustainable alternatives – again, delivered via digital means – is believed to lead to more 'greener' consumer behaviours (Demarque et al 2015).

Such an approach to digital technologies as, first and foremost, facilitators of information sharing and interactions with others is also central to the field of environmental communication. There, digital communication platforms (mainly, the Internet and social media) are often celebrated and instrumentalised as sites and tools of exchanging information – raising awareness of the climate crisis; finding like-minded allies; learning to argue and convince others; and engaging in on-line activism – while overlooking the actual material components and environmental tolls of these platforms (Tien Vu et al 2020; Gunster 2022; Baran and Stoltenberg 2023). Worryingly, the approach remains even in scholarship which focuses specifically on the materiality – and environmental toxicity – of the digital. For example, an edited collection titled *Sustainable Media* (Starosielski and Walker 2016), which firmly places digital communication within frameworks of resource extraction, wastage, and other forms of the slow devastation of digitisation, nevertheless celebrates the power of media as a 'means to come to terms with and help ameliorate the ecological harms produced by industrial processes' (2016: 3). Similarly, the contributors to *Carbon Capitalism and Communication: Confronting Climate Crisis* (Brevini and

Murdoch 2017), while noting the harms of communication, also emphasise its power to tell the untold 'back story' of the media industry and communication devices, as well as to address climate change and confront climate denial.

The second framework, prominent in sustainability studies and seen in other research and policy on digitisation, regards digital technologies as *tools* of sustainable innovation, eco- or carbon-efficiency, and efficiency more broadly. The ways digital technologies are understood as 'tools' – and indeed, the actual technologies that come under this framework – vary greatly, but all celebrate digitisation and naturalise it as the most suitable solution for sustainability. Some of the literature approaches digital and smart technologies as effective tools of resource management, including the management of energy consumption, water, waste, and recycling (see for example Mavropoulos, Tsakona, and Anthouli 2015; Di Salvo et al 2017; Venkatachary, Prasad, and Samikannu 2017). Key to this type of scholarship is the framework of efficiency (including 'eco-efficiency'), which is often linked to Big Data. Addressed explicitly and analysed in detail in some publications, and implied without elaboration in others, datafication is increasingly present in academic and policy conversations about digital technologies in the service of environmental goals: effective monitoring, analysis, and management of resources, it is argued, rests on having a large amount of data, which allows precision in understanding what has already happened, and effectively model future scenarios.

Another way in which the digital is rendered central to sustainability is modelling, referring firstly to manufacturing and construction, where doing things 'virtually' is seen as more sustainable because it saves actual resources (Kuntsman and Rattle 2019). An even more common use of modelling is linked to monitoring climate change and environmental damages. The examples of monitoring are multiple, and range from forestry, fishery, electricity, water use, and carbon footprint, as well as broader issues such as land ownership, or reporting of harms and ecological damages. More recently, academic and policy writing have turned to AI and its promises of data analysis and modelling of scenarios regarding climate change (Brevini 2021). Both earlier research on digitisation and more recent focus on AI rest on the assumption that digital capture (monitoring) and data analytics can provide precise and accurate analysis *and* 'democratise environmental governance' (Gale, Ascui, and Lovell 2017). Here, the data is seen as powerful due to it its scale and scope; but it is also transparent and impartial, allowing unmediated reporting from below; impartiality of analysis; and large-scale transparency. Another key assumption here is the promise of anticipatory knowledge (Weber 2019) which comes from predictive modelling – be it with regards to resources (their depletion, renewability, and so on) or climate events. The latter can be understood as part of the broader framework of anticipatory governance, where policy and state planning draw on technologically-mediated predictive analysis of actions and behaviours. At the heart of anticipatory governance is an investment in imagining and governing the future, which, as Muiderman et al have argued, is a 'core challenge for sustainability research and practice' (2020: 1).

In most of this scholarship, the digital is envisioned as environmentally neutral and devoid of its own footprints. In the rare discussions of digital harms, such footprints are either disregarded; or perceived as acceptable given the assumed environmental benefits of digital solutions. If concerns about digital harms are acknowledged, they are immediately followed by suggestions for mitigation – suggestions that often advocate for even more digital technologies, to fix the problems brought on by digitisation in the first place. A lack of explicit attention to environmental harms of digitisation is not merely an omission but a form of paradigmatic myopia (Kuntsman and Rattle 2019) towards material impacts and ecological harms of digital technologies, including those that appear to have been specifically developed for – and deployed in – environmentally oriented projects such as smart cities. There are a number of reasons for such myopia: from the overall framework of techno-optimism and techno-solutionism which automatically assign positive value to technological solutions; to more specific developments in *digital* solutionism (Kuntsman and Miyake 2022); from the magic of digital, to the politics of digital inevitability (Kuntsman 2020) – a worldview where futures are assumed to be necessarily digital, regardless of the perceived benefits and harms of digitisation; from the notion of the 'digital sublime' (Chen 2016) to the long-lasting tradition in digital/cyber studies that stubbornly imagines digital cultures as disembodied, de-territorialised, and immaterial despite a wealth of critique from feminist and decolonial scholars; from what Jennifer Good (2016) called 'symbolic annihilation' in popular and scientific representations that systematically obfuscate environmental harms of the digital economy; to the global financial and political power of this economy itself.

## Conclusion

In this chapter, we looked at the relations between smart cities, digitisation, and the environment, in order to de-centre the digital as the focus of conversation, and instead ask, how we can approach smart cities with the environment as our starting point. The chapter is in no way an exhaustive overview of all literature on smart, green, or sustainable urbanism, nor on digital sustainability – this was never our intention. Instead, our aim is to point to discontinuities and ruptures, and bring to the fore the mechanisms that obfuscate and conceal digital materialities and environmental harms of smart cities. More specifically, we drew on one example of lithium – a story one would not expect to find in a book on smart cities. While lithium batteries are needed for almost every device that is connected to any smart city's flows of data, information, and connectivity – consider the images of data flows in a smart city, as discussed earlier in the chapter – its mining in Chile or the labour of battery manufacturing in China both lie outside the imagined topographies and visualities of smart cities. In this chapter, we pointed to how the production of lithium batteries is contingent on the processes of resource extraction and exploitative labour, and embedded in colonial and imperial endeavours. And yet, these are excluded from how a smart city is imagined and visualised. In the rest of the book, we return to this organising logic which we describe as being *of out of sight, out of mind*, to better understand how and why the environmental tolls of smart cities disappear from view. Drawing on the conceptual frameworks detailed in the introduction, which range from communication to cultural studies to science and technology studies (STS) perspectives to political economy, cybercultures, and global injustices, we trace not only what we find in the stories of two smart cities – Helsinki and Manchester – but also what we don't find; what is not seen, and why.

More broadly, this chapter establishes our overall argument that how we tell the story of smart cities' environmental promises sets the stakes for how we understand, imagine, plan, and produce human, technological, and environmental relations on a broader scale. And in turn, how we arrange these relations forms a baseline for what assumptions are made about them, and about how we evaluate their implications for social, economic, and political conditions, and lived realities in the urban spaces. In other words, the way we tell the story of smart cities also shapes which environment, which people, and what needs and interests matter.

# TWO

# Helsinki, Kalasatama District

## Introduction

How to make an experimental concept seem real, convincing, and graspable? The videoblog 'What is a smart city?' (Nader Sayún 2020) featured on the website of the Smart Kalasatama district of Helsinki performs this act of persuasion. In the ten-part English-speaking vlog series Michel Nader Sayún, an MA student majoring in design, visits various smart infrastructures and projects in Kalasatama and conducts interviews with stakeholders. Each vlog delves into one aspect of Smart Kalasatama, including themes such as transportation, culture, citizen engagement, energy use and sustainability, and well-being. What is most interesting is the reference to the lack of knowledge, as well as the suspicion and scepticism, about smart cities that can be discerned in the vlog series.

In the vlog on the theme of energy use and sustainability, Michel visits the Hanasaari power plant, which is located in Kalasatama. Subsequently, during a Smart Kalasatama guided tour, Michel reflects on the paucity of knowledge regarding the energy infrastructure of smart cities. 'Apparently the program about sustainable energy is happening here, the guides know about it but they don't know much and they are not telling it to the people so nobody really knows what's going on' (Nader Sayún 2020: n.p.). In the same episode, Michel interviews Professor Eva Heiskanen from the University of Helsinki, who was involved in a pilot energy monitoring system in Kalasatama. Eva expresses reservations about the concept of smart city 'because it easily is sort of just focusing on digital solutions and big data and lots of collection of data' (Nader Sayún 2020: n.p.). 'Everybody who is developing smart cities should also read steampunk science fiction and think about other ways of being smart' (Nader Sayún 2020: n.p.), Eva says at the end of the interview with a smile and an assertive look that is telling of the simultaneous non-sensical and consequential nature of smart city projects.

Eva's concerns regarding the overreliance on digital solutions are not addressed or commented on in the entirety of the vlog series. Furthermore, although the involvement of residents is emphasised, the vlog series fail to identify how residents perceive the lack of knowledge about smart cities' environmental implications. This indicates that despite the concept of the smart city being poorly understood by the public and being viewed with suspicion by some experts, it is nevertheless heavily invested in, materially and affectively, as embodying the promise of a green digital transition. In this chapter, I examine the Smart Kalasatama district project in the Finnish capital Helsinki. The project began in 2013 and drew a close in June 2021. My analysis draws on a range of materials including documents such as the final report of the Smart Kalasatama district (Forum Virium Helsinki 2021), the report of the Six City Strategy Program (6Aika 2015), the textual and audiovisual materials found on websites such as the homepage of Smart Kalasatama, the website of the EU funding scheme, research articles on for example Helsinki's climate policy, energy transition, sustainability, and smart city practices, as well as my notes from exploring the Kalasatama area both online and offline.

This chapter proceeds as follows. First, I provide an account of my encounter with the district online and offline. Second, I contextualise the Smart Kalasatama district in terms of its role in the industrialisation and urban development of Helsinki, and the specific institutional and funding arrangements that shape the smart district project. Third, I zoom in on Kalasatama's environmental promises. I tease out how the environment is understood in these imaginaries and practices, what promises are made, and what human–digital–environmental relations are assumed and configured in these promises. Moreover, I ask what is elided in these promises, how they break, when and where. I then make explicit the various temporalities at work in the Kalasatama smart district project and lay bare their relations. I conclude this chapter by suggesting that the broken promises of smart cities in the context of Smart Kalasatama in effect render environment experimental in both spatial and temporal terms.

## Exploring the Smart Kalasatama district online and offline

Despite having lived in Helsinki for a considerable period and having passed through Kalasatama on numerous occasions, I did not know that Kalasatama is one of the pilot smart districts of Helsinki until I began preparing for this book project. In fact, it was only when I started researching smart city projects in Helsinki that I realised that the Jäätkäsaari area, where I had lived for more than two years, was also a pilot smart district. Additionally, I had used the vacuum waste disposal system, which I was previously unaware was an exemplar of smart infrastructure. I find this gap between the hailed significance and the lack of awareness of smart city projects perplexing. How is the concept of 'smart city' known and lived? For whom and in what way is it relevant?

Considering this felt dissonance, I decided to explore the Kalasatama district both online and offline. I began by examining the description of Kalasatama as a smart district. A quick Google search using the keywords 'smart city projects + Kalasatama + Helsinki' yielded about 69,500 results. The top results included Smart Kalasatama's homepage 'Fiksu Kalasatama' (Smart Kalasatama, Fin., Smart Kalasatama n.d.a), the website of the Nordic Smart City Network, as well as the webpage 'Fiksu Kaupunki' (smart city, Fin., Helsinki

Innovation Districts n.d.a) that provides information about all the smart city projects in Helsinki. 'Fiksu Kalasatama' and 'Fiksu Kaupunki' are both managed by the innovation company 'Forum Virium', owned by the City of Helsinki. I chose to begin by exploring the 'Fiksu Kalasatama' webpage as it provides extensive textual and audiovisual material, in Finnish and English, about the district. The curated content of the websites frames the district as smart, constructing its boundaries and landscape. In this sense, the website is not simply a representation but produces the coordinates of Smart Kalasatama, begging the question of where a Smart city begins and ends.

The front page of the 'Fiksu Kalasatama' website (see Figure 2.1) positions the district as a hub and a living laboratory, whose innovative and sustainable approaches are seen as exemplary not only in Helsinki, but of the Nordic area. The sense of a hub is made palpable in the image of the Isoisänsilta Bridge (Grandfather's Bridge) that provides the backdrop for the description of Kalasatama. The bridge was

Figure 2.1: Screenshot from Smart Kalasatama's homepage

Source: Fiksu Kalasatama/Smart Kalasatama, Forum Virium Helsinki (https://fiksukalasatama.fi/en/)

built in 2016, and links Kalasatama to the Helsinki mainland as well as the Mustikamaa island, a popular recreational area. It visualises the essence of a smart hub – a space for cooperation and collaboration between stakeholders such as residents, businesses, city officials, and researchers, and the integration of environmental sustainability and urban growth.

The liveliness of an urban innovation ecosystem, understood as a real, vibrant, and evolving entity, is temporalised through the rhythm of the self-scrolling images positioned at the centre of the front page. Automatically refreshed every few seconds, these images depict the four key features of Kalasatama. One of the key features is the figure of a hub described before. The other three are: 'Agile piloting', which is celebrated as the innovative model developed in Kalasatama. In this model, each pilot is funded for six months; 'Urban lab space', which is a co-working space located in the shopping centre REDI; and 'smart and resource wise district' that comprises the 'pioneer of Helsinki's climate goals' (Smart Kalasatama n.d.b).

Interestingly, although digital technologies are typically considered the central pillar of smart cities, they are not represented textually or visually as the key features of Kalasatama. The Internet of Things (IoT), including 'smart metering, smart parking, and shared electric vehicles', is only mentioned in one of the projects, the EU Horizon 2020 biotope project (whose homepage no longer exists despite a working link). The textual description of the IoT trials is accompanied by an image that foregrounds a smartphone held in one hand against the somewhat blurred backdrop of a loft apartment (see Figure 2.2). The smart screen displays the interface of a home automation app with various monitoring and automation functions. The image visualises a human–digital–environmental intimacy: the interface materialises as the intersection where the indoor and outdoor environments are sensed, categorised, monitored, and managed by a human subject through the digital eye. In contrast to the Isoisänsilta Bridge that evokes the sense of a hub that links rather than demarcates spaces, the IoT image

**Figure 2.2: Screenshot from Smart Kalasatama's homepage**

Source: Fiksu Kalasatama/Smart Kalasatama, Forum Virium Helsinki (https://fiksukalasatama.fi/en/)

generates the sense of a safe and controlled interior. However, the depiction of the proximate, intimate, and streamlined human–digital–environment relation made possible via the IoT trials tell us nothing about whether and how it works. In foregrounding the hand that holds the device, the image tries to convince the viewer that this is not an illusion or a magic trick, because the future is already in your hands.

In comparison with the four key features of Kalasatama, the IoT projects appear to be ad hoc. While the key features point to the essence of Kalasatama as an 'open innovation ecosystem' (Smart Kalasatama n.d.b) that offers the conditions for innovation projects, the question remains as to whether and how these projects are integrated, inter-dependent, and co-evolving – the function of an 'ecosystem'. The curation of 'Fiksu Kalasatama' webpage establishes connections between the social (for example, the feature of Agile piloting underscores community building and stakeholder involvement), economic (such as the feature of Kalasatama urban lab foregrounds the profitability of experimentation and innovation), and environmental (for example, the feature of a smart and resource-wise district focuses on issues including climate change and waste disposal) elements of Kalasatama. It produces an imaginary of a holistic and smart system.

The question is whether and how the different features and projects integrate into a smart ecosystem, and how is it experienced in everyday lived practices? In spring of 2023, I used the Smart Kalasatama map on the project's website as a reference point for exploring Kalasatama on foot. Initially, my plan was to take a guided walking tour promoted both on the webpage and in the final report of Smart Kalasatama. The guided tours are organised by Forum Virum Helsinki and outsourced to freelancers who charge approximately 200 euros for a two-hour walk – a cost that is hardly affordable for a visitor or a resident, speaking volumes about the ways in which the smart district functions as a brand and a marketing strategy for the City of Helsinki. Instead of paying for the guided tour, I opted for exploring the district myself, by comparing the smart district map and the project description with my own experiences of the area.

To get a better sense of what constitutes the boundary that demarcates the Smart Kalasatama district from the other 'not-so-smart' neighbourhoods, I decided to walk to Kalasatama from Helsinki train station. The Hanasaari power plant and the high-rise buildings were easily spotted from far away. On the day of my visit, the power plant, which was due to close on 1 April 2023 was still operating. The steam rising from one of the power station's chimneys resembled clouds. Their white colour and fluffy shape contrasted sharply with the stiffness and heaviness of the red-brick power station. Equally stiff, but with a much more modern look, were the high-rise buildings. Their silk-screened glass façade glistened in the sunlight. A tower crane near the buildings attested to the ongoing construction of new high-rise buildings. The power plant and the high-rise buildings – the two important landmarks of Helsiki – testified to the changes in the city's urban landscape and energy infrastructure.

Despite the good weather, the area looked grey and concrete. The air felt dusty. The bodily experiences of the environment created dissonance with the depiction of Smart Kalasatama as 'showcasing Smart'n'Clean' Helsinki with 'water and air' being one of the four areas of focus (Smart Kalasatama n.d.c). I deliberately looked for and tried to locate the smart environmental infrastructures described on the map. However, besides the vacuum waste disposal system that was made available in some of the residential areas, I could not find much. The more I tried to look for the environment through the lens of smart in Kalasatama, the more it seemed unlocatable; the more I attempted to make visible the relation between the digital and the environment, the more I wondered *what, where* and perhaps even *when* exactly this environment was.

The high-rise buildings are considered to showcase the state-of-the-art in energy-efficient design and provide the testing grounds for IoT trials. In fact, the installation of home automation system, including hardware such as 'relays, meters, and communication capable controllers' (Härkönen et al 2022: 2) in all new apartments (approximately 10,000 apartments by 2035), was made a condition of land transfer to development and construction companies in Kalasatama by the City of Helsinki. It was of course impossible to see the home automation system from the outside of the building, let alone understand and evaluate their function and impact on the environment – such an exploration would be particularly interesting, although it lies beyond the scope of this book. What could be seen, however, was a site where construction waste was scattered around close to the REDI shopping centre (see Figure 2.3) and concrete blocks that were piled up next to the road (see Figure 2.4). I will return to the home automation issue later in this chapter.

To get a better sense of the role of green spaces and the environment in the smart city, I tried to identify the location of the 'Healthy Liveable Neighbourhoods' project featured in the news column of the front page of 'Fiksu Kalasatama'. In

Figure 2.3: The site of construction waste

Figure 2.4: The concrete piled up by the road

**Figure 2.5:** Screenshot of the 'healthy liveable neighbourhoods' video that introduces the Green Kalasatama app

Source: Kalasatama, Forum Virium Helsinki and Nordic Smart City Network (https://fiksukaupunki.fi/en/projects/green-kalasatama/)

the video that introduces the Green Kalasatama app that is a key element of the project, residents are shown looking at their surroundings through the lens of an AR (augmented reality) app on a tablet (see Figure 2.5). They place trees, bushes, and plants into their desired spaces in the virtual environment, making the area appear green and flourishing. Walking in the areas depicted in the video, I could not see the planted trees, bushes, or plants. In fact, Kalasatama, as I saw it, lacked vegetation entirely. Trees are sparsely planted (see Figure 2.6). As Vierkko et al (2022: 187) observe in their study on the Kalasatama district's SMARTer Greener Cities (2020–2023) project, 'although the recently established public park is centrally located and accessible for all residents, the overall scenery of public areas is filled with hardscapes: paved areas, walk and driveways with arbitrary street trees, and walls of high-rise buildings'. What then constitutes the environment in the smart district? To address this question, I turn next to contextualise Kalasatama through its energy and environment specificities and institutional and funding arrangements.

Figure 2.6: The park

## Kalasatama: from the engine of industrial development to the district of creative innovation

Kalasatama, whose literal English translation is 'fish port', is a well-known area of Helsinki. The once dingy harbour landscape is a key site for the acclaimed Aki Kaurismäki's 2002 film 'A Man without A Past'. In the film, the protagonist, who lost his memory after being beaten and mugged by hoodlums, finds a new home among a community of outcasts living in the abandoned shipping containers in Kalasatama. The area is also known for its obsolete industrial spaces, which are being transformed into a place of cultural, technological, and entrepreneurial innovation. For example, the Suvilahti area of Kalasatama, once an industrial premise, is now a cultural centre annually hosting the popular urban art and music festival, 'Flow'. Kalasatama is also home to some of Helsinki's most important industrial and architectural landmarks. The Hanasaari B coal-fired power plant, completed in 1976 and decommissioned in 2023, was an important energy infrastructure that produced electricity and district heating. It was also 'the third-largest industrial emission source in Finland' (City of Helsinki 2020), generating 2 per cent of the country' total emission. The newly built high-rise buildings (fifth out of the eight planned buildings being constructed at the time of writing) in the centre of the Kalasatama district are the tallest buildings not only in Helsinki but in the entirety of Finland.

The Smart Kalasatama project is often told as a story of transformation where an abandoned harbour and industrial wasteland is rebuilt into an area of innovation and smart living. Kalasatama is one of the three districts of the old town bay of Helsinki. The construction of the area began in 1825, as factories and workshops were ordered to be located on the outskirts of the Helsinki city due to the risk of fire. In the late 19th century, this area saw a rapid population growth due to the internal migration of workers from nearby municipalities. Energy infrastructures such as Helsinki's first electricity plant and gas plants were built in this district in the early 1900s. By the end of 2008, the port was moved from Kalasatama, leaving only a few small-scale industries operating in the area. In 2009, Helsinki city council began to rebuild Kalasatama from a brown field to a new district of creativity and innovation – one of the largest urban development projects of the City of Helsinki.

The transformation of Kalasatama involves not simply infrastructural rearrangements. It is a multi-layered socio-political, economic, and environmental process. The imaginary and narrative of transformation – from a wasteland and a site of industrial remains to a cultural, creative, and innovative centre – justifies the usage of the area for smart experiments, which in turn allows for the maximisation and extraction of land rent (see Andreucci et al 2017). The City of Helsinki is both the owner of land and the regulator that holds decision-making power over Kalasatama's development. While public planning and strict land use regulations were the standard approach of the City of Helsinki, in recent years it has begun to outsource urban planning to private developers. For example, in June 2020, the city reserved a plot of land in Kalasatama for Suvilahti Event Hub, which is 'a real-estate development project backed by private investment' (Hyötyläinen 2023: 220). It rebranded Kalasatama as a new cultural and innovation centre to raise its value, and demolished public facilities, such as the popular skatepark built by skater communities, for commercial use, constructing offices, hotels, and high-end residential buildings that generate high rent revenue (see Hyötyläinen 2023).

The rebuilding of the area as a residential and business district also entails spatial rearrangements and environmental transformations. Many of the new residential buildings are constructed on the landfill that was used to fill the Kalasatama shoreline and to extend and connect it to the mainland. The process began in the 1930s and continued until 1980s. The landfill consists of surplus construction material such as construction waste. According to Helsinki city's environmental report of 2014, multiple sites of soil were contaminated by the landfill and excavated, treated, and backfilled (City of Helsinki 2020). The new public park of Kalasatama is built on top of the flat landfill: it is worth mentioning here that the public park of Jäätkäsaari – the other post-industrial harbour area of Helsinki that has been reconstructed as a smart district – is built on an area of the most contaminated soil, 'with the ground unfit for housing for that reason' (Ameel 2021: 104).

The narrative of transforming post-industrial wasteland affectively draws on a heroic tale of human transformation of nature, understood as an empty space of wilderness. According to Lieven Ameel, such narrative reproduces a 'settler rhetoric' where nature, understood as 'a spatial tabula rasa', awaits 'mankind's guidance' (2021: 59). However, the environmental history of Kalasatama is neither empty nor natural, in the sense of being unspoilt and unaffected by human activities. The contaminated soil and the man-made shoreline have developed specific ecological system dynamics over time (see for example Vierikko et al 2022), which are affected by the construction of new infrastructure on these sites. The sedimented environmental processes that are still slowly morphing and unfolding are sidelined in the narrative of smart transformation. This could be partly explained by whether and how the funding schemes frame environmental issues.

The Smart Kalasatama district project is part of the 'The Six City Strategy – Open and Smart Services' programme that was approved by the Finnish Ministry of Employment and the Economy in June 2014, and ran during the period 2015–2017. The programme was carried out in the six largest cities in Finland: Helsinki, Espoo, Vantaa, Tampere, Turku, and Oulu. The focus areas of the programme included open data and interfaces, open participation and customership, and open innovation platforms. The programme had an experimental focus, consisting of 'spearhead projects', 'practical pilots', and 'trials', for example, whose results include the 'standardisation' and the possibility of 'duplication of functional solutions and innovations' (6Aika 2015). As its description reads:

> The primary objective of the strategy is to strengthen Finland's competitiveness by using the country's six largest cities as innovation development and experimentation environments, with the aim of creating new know-how … [It] offers new opportunities for a number of operators, including ICT companies, geographic information service companies, the security sector, developers of traffic services, developers of detector and sensor technologies, suppliers of clean- and greentech solutions, the media sector and other creative sectors as well as various service sectors. (6Aika 2015: 5)

The funding for the 'The Six City Strategy' came from the European Regional Development Fund. The theme of the funding period 2014–2020 was 'innovation and research, the digital agenda, support for small and medium-size enterprises (SMEs), the low-carbon economy' (European Commission n.d.). The requirement for allocating funding for these thematic projects varied depending on the 'developed-ness' of these regions. For example, whereas for more developed regions, a minimum of 80 per cent of funds must focus on at least two of these thematic areas, for less developed areas 50 per cent of funds were required. While the channelling of funding to low-carbon economy projects was compulsory in all regions, 20 per cent of fund allocation was required for more developed regions and 12 per cent for less developed regions. In many ways, the emphasis on digital-orientated innovation and entrepreneurship echoes the key features of smart cities. This becomes even more evident in the current funding round of 2021–2027. The first of the five funding priorities are 'more competitive and smarter, through innovation and support to small and medium-sized businesses, as well as digitisation and digital connectivity' (European Commission n.d.).

While the funding body provides the guidelines, the decision-making power is passed on to each member state. In the Finnish context, the funding is managed by various organisations. The Six Cities programme's joint management group, consisting of directors in charge of the business and innovation matters, supervise and monitor the implementation of the funding scheme. In Helsinki, Forum Virium functions as an intermediary that coordinates 'companies, universities, other public sector organisations and Helsinki residents' (Forum Virium Helsinki n.d.). It plays a determining role in shaping the direction of the smart city projects. For example, Forum Virium initiated the Virtual Verdure Project, that ran from September 2019 to April 2020 and whose environmental objectives were not explicitly stated in the funding scheme. It partnered with WSP (a Canadian consulting firm providing, for example, engineering and design services) and InnoGreen (a Finnish greenery service company whose products include green walls and decoration for indoor spaces and garden designs for urban spaces). This project was used to mark the environmental focus of Kalasatama, a feature that was said to differentiate the Smart Kalasatama project from other smart district projects in Helsinki. However, only the northern side of Kalasatama was used as a testing ground for the Virtual Verdure Project. As I will show next, Kalastama's environment and energy projects are fragmented, disjointed, and are often limited in scale and duration.

## Kalasatama and its promises

### The promise of becoming green and sustainable

In comparison with smart districts such as Jätkäsaari that invest in mobility projects that utilise digital technologies to make traffic systems smoother and safer, the focus of Kalasatama is on smart green infrastructures that support Finland's climate goals. As the website reads:

> The intelligent energy systems project by Helen, ABB and Fingrid works on Kalasatama's smart grid and its co-products, such as electric car network and energy storage. The new vacuum-based waste removal system, Imu, is ready operational. Kalasatama – just like the rest of Helsinki – is investing heavily in opening up public data and making the most of it. The aim is to utilize the data in city services, for example to provide information about local air quality or car-sharing options nearby. (Smart Kalasatama n.d.c)

During my visit, major construction projects were still being carried out. Much of the construction took place around REDI, a shopping and transportation complex. REDI was awarded LEED (Leadership in Energy and Environmental Design) platinum certification in 2019, an environment-efficiency measuring standard that originated in the US. According to SRV – the developer of REDI – the sustainability features of REDI include water-saving fittings, waste management, and low emission surface materials, for example. Its eight-tower buildings – residential apartments – are also smart, incorporating features such as smart meters that measure and collect data about residents' electricity and water consumption. However, as an expert on the energy emissions of buildings told me (personal communication), the environmental impact of the entire life-cycle of buildings – the extraction, production, transportation, and manufacturing of construction materials that are used to construct, maintain, retrofit, and demolish a building – is often overlooked. Ironically, the demolition or retrofitting of old, energy-inefficient buildings are themselves an energy-intensive process. Their energy consumption is difficult to measure; and their footprint is rarely mentioned in reports. The measurement of the environmental efficiency of REDI and the high-rise buildings does not take into account the environmental impact – the usage of water, greenhouse gas emissions, as well as other energy and resource use – of the production, transportation, and usage of materials like concrete that high-rise buildings heavily rely on (see for example Pomponi et al 2021). Moreover, the construction waste, containing toxic substances and microplastics, risks contaminating soil and water over time (see for example Salkkonen 2013; Hird 2022).

Besides these hidden environmental impacts, the actual applicability and efficiency of the home automation system is hardly ever mentioned. As discussed earlier, as a condition of land transfer, the City of Helsinki requires the installation of a home automation system in all apartments in Kalasatama. The automation system consists of hardware and software such as panel boards inside each apartment and in the central control of the buildings, sensors for monitoring indoor temperature, interfaces for managing the automation system, and servers that collect meter and sensor data. The server can also be used for external data communication and transfer. The main purpose of a home automation system is to manage the balance between demand and supply on the electricity grid and reduce peak energy usage. In arranging the demand load of various domestic activities – lighting, cooking, laundry, and so on – the automation system shifts the energy usage from a paradigm where the supply follows the demand to a demand-response model where the demand is determined by when the supply is made available (see for example Barsanti et al 2020).

However, the installation of home automation system does not *automatically* translate into its applicability or more effective energy management. Ideally, a set of other conditions need to be in place. All the home appliances should be smart, and their data integrated and communicated externally via the server (see for example Barsanti et al 2020). Residents need to understand how the system works and have the willingness to change their energy usage practices. In other words, and more fundamentally, the effectiveness of a home automation system depends on whether and how it is put into use. In the case of Kalasatama, even though the installation of such a system is required, who and what is responsible for its application and management is unspecified. As Kalevi Härkönen et al point out in their research on the usage of home automation systems in Kalasatama:

> Installation of home automation belongs to electrical contractors, but after the installation on a site is complete and people have moved into their apartments, their work is over. Energy service and building management companies operate with HVAC systems of public and commercial buildings and with professional customers, not with residential electrical installations and private customers. This may have contributed to the low utilisation of home automation since apartments in the district have begun to complete. (2022: 14)

Another important concern that is often elided in smart home discussions is how energy is produced in the first place. Kalasatama utilises district heating and cooling provided by the energy company Helen, a private enterprise owned by the

City of Helsinki (Helen Ltd n.d.). Helen has started to renew and decommission coal-fired power plants and transition to 'renewable energy' such as wood biomass for district heating. However, what counts as renewable is subject to debate. As Tere Vadén et al (2019) point out, the emissions from burning biomass and logging existing forests – carbon sinks – are in fact higher than emissions from fossil fuels. Moreover, the biomass industry harvests mature trees that are considered valueless for the forest industry, but are significant for biodiversity, soil, and water quality (Speare-Cole 2021). The smart solutions might succeed in collecting more data (whether or not it is actually useful) or even changing some energy usage practices. But their implication for larger-scale and long-term transformations of energy infrastructure is limited.

Recall the 'Green Kalasatama' project described earlier, featuring the AR app which allows for visualising vegetations and green spaces that *do not exist*. The project was carried out in the summer and autumn of 2021, when residents were invited to participate in imagining how Kalasatama *could* look like by planting digital plants on the app. This practice of co-creation and co-development is considered a cornerstone of Kalasatama's long-term vision as a green district. As the project's final report reads: 'a greener city requires new and long-term design for which a wide range of expertise is needed. Kalasatama has served as a test area for co-development and smart solutions' (Forum Virium Helsinki 2021: 18). However, the aftermath of this project, both in terms of the data generated and the implications for the planning and construction of green spaces in Kalasatama, remains unclear. The project acknowledges this limitation by stating that '[i]n the future, the use of such data remains to be explored in detail after the early phase pilot' (Helsinki Innovation Districts n.d.b). What happens to the data, the AR apps, the knowledge, labour, and the funding used to produce it? Can the opinions of the residents, so valued by 'Green Kalasatama' project in particular, and the Smart Kalasatama district in general, actually have any effect on the planning and decision-making process? And when, and how, can the virtual green space translate into a physical one? These questions do not seem to be of concern for the project. The low usefulness of its data is not seen as a sign of failure, however. Rather, it appears to be a normalised and expected feature of a long-term approach that relies on technological advancement. This is because, as the project description reads, 'There is more to learn as the technologies develop and users' own devices become more advanced' (Helsinki Innovation Districts n.d.b).

My observation resonates with Kati Vierikko et al's (2022) study on Smart Kalasatama's Virtual Verdure project. Situated in the field of urban ecology studies, Vierikko et al's (2022) research shows that the co-creation methods are not integrated in the planning of the area. The dominance of technical optimisation hinders 'the comprehensive enhancement of urban biodiversity and integration of ecological systems into social-technical systems of urban infrastructure' (Vierikko et al 2022: 188). 'Green Kalasatama' is a good example of smart city projects that deploy digital visualisation of the environment – the green elements and spots on screen – as standing in for the environment, and as signs of environmental solutions. In other words, the form, content, logic, and practice of digital technologies shape and constitute environmental problems and solutions as *digital* ones. In this logic, the production of data is already a solution in itself, even if this is what I call passive data. In the context of mobile apps, passive data can mean the reception of data when an app is running in the background. Here, I refer to passive data in two inter-related senses. First, it is data that is updated and made accessible but is not engaged with, or even noticed. Second, passive data does not have any implication for systemic, structural, or environmental changes. Such is the case of data generated by the Green Kalasatama AR app.

Another good example of passive data is the air quality map displayed on screens in public transport in Helsinki, including the metros that pass through Kalasatama REDI shopping centre. The screen displayed various advertisements, news, and information of things such as traffic, weather, and air quality. Each display lasted a few seconds. Interestingly, there was a glitch in the display. The bottom half of the air quality index map was not shown. However, the passengers on the metro did not seem to notice the glitch at all. Most of them were busy checking their phones. Some were staring out the window. Some were talking to other passengers. Others appeared to be resting with their eyes closed. The air quality data was passive, as it was irrelevant and inconsequential to the residents who were supposed to benefit from it. It did not seem to be linked to structural change concerning energy transition, environmental pollution, or biodiversity loss.

Despite the promise of a 'holistic' approach, in Kalasatama there is a separation between green infrastructure and smart infrastructure. In its final report, 'smart infrastructures' and 'green infrastructures' are categorised as two different kinds of features. It is said that smart infrastructures afford energy efficiency and support 'a more ecological carbon-neutral lifestyle' (Forum Virium Helsinki 2021: 2), whereas green infrastructures integrate green spaces in the city. Interestingly, while the link between climate change and environmental issues is recognised, their solutions are separate. This discrepancy cannot be explained simply in terms of a different focus on technology and the environment, especially given the heavy investment in and reliance on digital technologies to solve environmental problems. As I will show in the next section, this fragmentation and disintegration is configured by and takes shape as the multiple temporalities of smart cities.

### The promise of 'one more hour a day'

The temporality of Kalasatama smart district is multi-layered, fragmented, and, in many cases, self-contradictory. Its most visible temporal element is the promise of 'one more hour a

**Figure 2.7: Screenshot from the Forum Virium homepage**

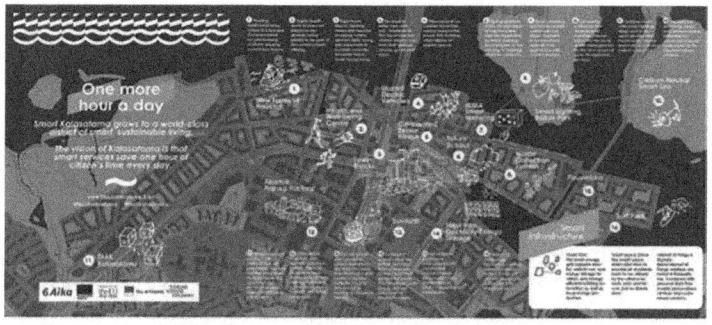

Source: Fiksu Kalasatama/Smart Kalasatama, Forum Virium Helsinki (https://foru mvirium.fi/en/projects/kalasatama-smart-city-district-of-helsinki)

day' (see Figure 2.7) for its residents that is mentioned earlier. It has attracted international media attention and elevated Helsinki to the position of the second-best smart city in the world according to the 2020 smart city index (IMD 2020). The 'one more hour a day' slogan captures the two aspects of the temporal promises of smart cities.

The first concerns efficiency and increased mobility because of a wide range of smart practices – from better coordination of, and information about transportation, to experimentations such as using robots to deliver food from supermarkets to apartments. One of the pilot projects uses robot buses that 'serve the residents as a solution for moving the last mile' for example 'between the REDI shopping center and isoisänsilta bridge' (Forum Virium Helsinki 2021: 15). According to Google Maps, the distance between REDI shopping centre and Isoisänsilta Bridge is about 800 metres, and should take about 10 minutes on foot. Given such a short distance, it is unclear why a robot bus would be needed, especially considering the good condition of the pedestrian path and designated bike lanes, seen as important features of Smart Kalasatama, as stated in the final report:

More and more attention has been paid to the different factors of residents' well-being in urban areas following positive research results on the impact of green environments on health, and learning how the urban infrastructure can contribute to people's active and mobile lifestyles, for example. In the construction of Kalasatama and the agile pilots carried out in the district, well-being in an urban environment has been an important approach. Kalasatama is a densely-built residential area with good access to green recreation areas as well as city activities and services that can be reached on foot. (Forum Virium Helsinki 2021: 20)

The report highlights the importance of active lifestyles and the affordance of green urban infrastructure. For example, images of happy looking residents cycling are featured both on the cover of the report (see Figure 2.8) and in the section describing 'everyday well-being' as one of the four smart

Figure 2.8: Screenshot from the Smart Kalasatama final report cover

Source: Fiksu Kalasatama/Smart Kalasatama, Forum Virium Helsinki and Nordic Smart City Network (Forum Virium Helsinki 2021: 1)

features of Kalasatama district. Given this, the implementation of robot buses and robot services, however short the duration, is perplexing. Who is such service aimed at? What kind of need is it supposed to meet? Is it only accessible to residents who utilise the smart services in Kalasatama? What is the relation between the smart time of one more hour a day, and other temporalities such as the slow contamination of soil from construction waste, the formation of new ecological dynamics, or the short duration of project funding for pilots and trials?

### The promise of long-term development

The second temporal promise of smart cities is that of sustainability, understood of as long-term *development*. The economic and managerial logic of sustainability has since its inception concerned not so much the well-being of nature itself but how to best manage its resource to sustain economic development. In the case of Kalasatama, it could be said that the efficiency of smart living posits time as a form of resource for a productive life, as well as for managing energy and natural resources. For example, the report describes a smart housing pilot that uses the Internet of Things as follows: 'A key theme in smart housing pilots has been how smart solutions can help city residents in everyday life, how they can make daily and weekly activities easier, and save not only time but also other resources such as energy and natural resources' (Forum Virium Helsinki 2021: 14).

The temporal promises of longer-term development and an easier everyday life are intertwined with other temporalities, such as those of an experiment, which, in turn, is shaped by the funding schemes, their life cycles, and the organisations involved in running them. Here is one telling example of such relations: in 2013, Forum Virium shifted the focus of Kalasatama from key infrastructural changes that are bigger in scale and longer in duration to shorter and smaller-scale experimental pilot projects. As Kaisa Matschoss and Eva Heiskanen explain:

> The original focus of the experiment was on developing business models based on urban smart grids utilizing experimental infrastructure for smart metering and control. The incumbent energy company has also had ambitions in developing solar power, district cooling and energy storage. Since the innovation intermediary, Forum Virium Helsinki, was engaged in 2013. The focus has turned more to smart living, including intensified co-development of services (open data, transport and sharing economy) together with users and startup businesses. (2017: 88)

Interestingly, the final report does not deny or omit the short duration of pilot projects and their seemingly limited shelf life. Instead, the report justifies, values, and celebrates it. The report presents the short duration of an experiment as a necessity for long-term success, as it allows for generating and testing more data and solutions. It is through trial and error that the best solution can be found and implemented. Here, the long-term goal – for a city to become sustainable and smart – justifies the short–term-ness of a project economy. This, in turn, gives ground for the vagueness of the long-term goals – what should be achieved and by when, and what is needed to get there. Ironically and crucially for our discussion, given that there are always more experiments to be conducted, better and smarter solutions to be found, and more data to be collected, the long-term goals can be endlessly postponed.

## Conclusion: experimental temporalities and the environment

In this chapter, I analysed the Smart Kalasatama district and its promises, focusing specifically on the environment.

I began by exploring Kalasatama's online representations and offline, physical spaces. I analysed how the webpage 'Fiksu Kalasatama' constructs the spatio-temporal boundary and landscape of Kalasatama as smart, and asked how such smartness might be experienced. To better understand the configuration of human-technological-environment relations, I contextualised Kalasatama's environmental promises in the district's energy and environmental histories and financial and institutional arrangements. In so doing, I demonstrated the inter-dependencies and discrepancies of Kalasatama's multiple temporalities.

By way of conclusion, I elaborate on how the temporalities of the digital constitute and encapsulate the broken promises of smart cities. As I have shown, the solutions that digital technologies may provide are limited in scale and duration for the massive infrastructural changes that are required to fulfil bigger promises such as green transition. Despite the assumed centrality of digital technologies in smart city visions, the actual urban development practices of smart cities still predominantly concern changes in land use and infrastructures such as residential buildings and energy systems. The celebration of the potential of digital technologies' affordances to improve energy efficiency obscures the crucial question of how and from what sources the energy that powers the digital is produced? The temporal promise of the digital – real-time measurement, agility, and flexibility (new software and programmes can be updated, new data can be generated) – also allows the long-term temporality of major changes, such as energy transitions, to be ignored.

For example, Kalasatama's implementation of energy and district heating from biomass creates technological and economic path-dependencies and lock-ins that could result in 'yet another decades-long increase in carbon dioxide emissions' (Váden et al 2019: 7). As Váden et al explain, lock-ins and path-dependencies happen when 'there are high infrastructure costs or there exist legacies of previous infrastructure that would be expensive to change or retrofit' (2019: 2). Biomass is seen as an easy replacement for coal when generating district heating. However, this is not because it is supposedly 'renewable'. Rather, burning biomass does not require changing the existing district heating system, which is built for certain temperatures and pressure. In turn, the investment in biomass burning will lock-in the material, technological, economic, and political path-dependencies to these high-emission systems for decades. Moreover, the life-cycle of the forest means that felling mature trees results in the loss of carbon sinks and changes to ecosystems that have long-term implications. In addition to the temporality of energy infrastructures, the digital is shaped by various social and environmental processes, such as an end to a project and/or to funding. For example, the 'news' section of the Smart Kalasatama website was last updated in November 2021, suggesting that following the conclusion of the project, the digital space of the project was no longer maintained. This indicates that the digital representation of the district is now frozen in time, and the 'real-timeness' of the digital city is now disabled, unavailable, and irrelevant. Unlike many of the smart project websites which are taken down after the projects ended (as described, for example, in the next chapter), the Kalasatama project website is preserved.

In that respect, the once 'real-time' project can now be seen as a digital ruin. Vincent Miller and Gonzalo C. Garcia define digital ruins as '*online spaces that have been largely abandoned by their users but continue to exist intact*' (2019: 437, emphasis in original). My understanding and usage of the term here draws inspiration, and also differs, from Miller's and Garcia's, as it is unclear who the intended users of the site were in the first place, let alone whether and how they might still engage with the website once it is no longer being updated. The website is not a virtual construction of spaces such as buildings or rooms, or streets. And much of its content does not concern space in a self-evident or narrow sense. Nevertheless, it spatialises. It maps the pilots, trials, and infrastructures as parts of an allegedly integrated smart district. In this way, it flattens out their differences in terms of scale, duration, function, material conditions and environmental impact and makes them all equally smart.

And it does so through multiple temporal expressions, too. Liveliness – the displayed self-scroll images on the front page; temporal stasis – an extended presence of the digital where time seems to have stopped unfolding; temporal dissonance – as links to external websites still work even as the websites themselves disappear; and ellipses – begging the question of what happens to the various pilots, trials and experiments, what fails to be updated, what has been left out of these descriptions and what comes after. This specific temporality of digital ruination becomes felt and intelligible as a form of brokenness – the incommensurability between the speed of digital production, its disposability and the temporality of the non-digital – that is nevertheless what sustains the 'real-time' promise of the digital logic.

# THREE

# Manchester

## Introduction

Manchester is among several UK cities which have been developing and adopting various smart city initiatives since 2010s, building on the legacy of digital strategies published in 2008 and 2012 (Cowley and Capriotti 2019). Named as one of the lead UK smart cities, Manchester's innovation plans and policies have been documented and analysed as smart city 'case studies' across several disciplines, from urban planning and development to geography, citizenship, ethnography, and art (Caird 2018; Cowley, Joss, and Dayot 2018; Fraser and Willmott 2020). In academic publications, Manchester is often presented as part of the larger processes of urban digitisation, urban governance, and 'techno-publics' (Cowley, Joss, and Dayot 2018), that take pace the UK, in Europe, and globally. It is also noted that, for Manchester, becoming a smart city is part of a longer trajectory of industrial modernity, transforming from early industrial, to post-industrial, to hyper technological (Fraser and Willmott 2020). Policy narratives and media essays often depict Manchester as a 'city of pioneering innovation', of which the smart city is only the latest stage (Collier 2019; Slatcher 2016). See, for example, the infographic presenting the Manchester context of smart city developments in one of the reports compiled by the Manchester City Council's long-term expert on smart cities, Adrian Slatcher (Figure 3.1, from Slatcher 2016). The infographics link industrialisation and transport development to nuclear science, computing, and the

Figure 3.1: Manchester context

**Manchester Context**

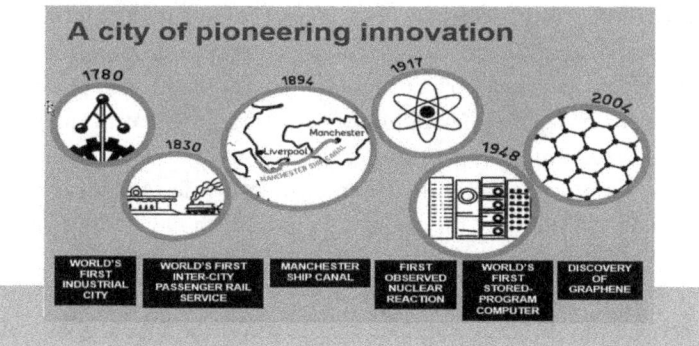

Source: Slatcher (2016)

discovery of graphene, all of which took place in Manchester. In a similar vein, a media report, written by a Manchester University research fellow, links smart city developments to the city's history as 'the birthplace of the (first) Industrial Revolution, where the steam engine first roared to life, where the atom was split, where "Baby", the first stored program computer was built, all of which changed the modern world forever' (Collier 2019: n.p.).

In these and many similar narratives, the digitisation and smartification of today are tied to these centuries-long, and seemingly uninterrupted, lines of scientific progress. Major disruptions such as wars, decolonisation, or political upheavals, as well as more routine unsettling of economic reforms and crises, changes in governance, migration, citizen protest, and periods of harshening austerity, are entirely absent in this genealogy. Instead, there is a sense of continuity of innovation and discovery, a continuous link from the industrial revolution of the 18th and 19th century to the digital revolution of today. What's more, the latest period between the beginning of smart city initiatives in Manchester (and elsewhere) in the 2010s and the time of writing this book in the early 2020s is sandwiched between two major historical events – Brexit and the COVID-19 pandemic – and several other pivotal moments, both local and global. These range from the 'devolution' process – a delegation of financial and political power from the UK's national government to regions and cities across the country, giving Manchester Council more autonomy in decision making around the city development – to ongoing austerity politics and most recently, at the time of writing this book in 2022–2023, the energy and cost of living crisis. These, too, are almost entirely absent.

What is similarly absent in these narratives of continuous innovation in the city are environmental questions and problems, both past and present: the city's actual geography and landscape; the region it is part of; the socio-demographic inequalities of its different neighbourhoods; and the history of environmental concerns, from the extensive pollution of industrialisation and de-industrialistion, to post-industrial urban decay, high traffic and low air quality, and inequality in accessing green spaces (Fraser and Willmott, 2020). Here, it is worth noting that despite the *current* discursive presence of climate change and the environment in policy documents and the public imagination, it has been decades since climate change 'first took center stage' in Manchester (Knox 2020: 35), after years of muted and suppressed climate policy; and even then it 'took so long for the city to think [...] about climate change as a core consideration of urban politics' (Knox 2020: 38).

This chapter goes against the grain of both the long history of silencing environmental issues in city planning and vision, and the current hype of digital innovation by asking: what would happen, if instead of centring the discussion of Manchester as a smart city on a genealogy of scientific pioneering and technological solutionism, we centre it on environmental questions and concerns? How and where can we look for the signs – or absences – of environment in the narratives and materialities of smart cities? How do we track smart city legacies through the lens of environmental care, nature, and climate change? What do these legacies grow into? The task is simultaneously obvious and elusive, for any smart city – and Manchester is in no way an exception – is both a discursive construction, a dream, a ghostly digital trace, *and* a material reality.

This chapter is in not an exhaustive description of what Manchester as a smart city *is*, nor is it a comprehensive assessment of its digital innovation successes and failures.[1] Rather, it is a snapshot of a current moment in the life of a smart city – a snapshot that is also deeply connected to the sense of past and future, historicity and temporality, as discussed later in this chapter and as the Conclusion. This chapter is a brief visit into Manchester's smart city imaginaries and materialities, their visible and invisible environmental impacts, and the ways these are understood, today, in relation to digital technologies. Methodologically, the chapter has several entry points: documentary research, interviews, and snippets of auto-ethnography. As a digital scholar, my work focused primarily on visual and textual materials about the city, available almost exclusively in a digital format, and mostly located on the internet. My exploration consisted of collecting and mapping policy documents, industry reports, news and social media posts, and academic papers about the city's current digital strategy, as well as the two major smart city projects that were launched in the mid-2010s and completed just before the COVID-19 pandemic. To get a further sense of how Manchester as a smart city is seen in relation to environmental concerns, and how digitisation and the environment are understood in relation to each other, I have interviewed five individuals from several city- or region-based organisations, who work either with digital technologies, or climate and environmental issues.[2] These interviews, too, were digitally mediated: although all the interviewees were based locally, as was I, we spoke on Zoom, in a post-pandemic habit where videoconferencing became both safe and convenient for many people, working from home or navigating complex schedules. At the same time, my exploration of the city was physical and embodied. As a resident of Manchester, I also live, work, and commute around the city on a regular basis, navigating it by bike, foot, or public transport. My accounts of movements across the city are included in the chapter. Conceptually, the chapter is driven by the questions that inspire this book: What are the environmental promises of smart cities? Where and when do they begin and end? What happens when these promises break, and what is left in their aftermath?

## Smart city demonstrators

In the mid–2010s, two major smart city projects took off in Manchester. One was called City Verve, launched in 2016 and led by a consortium which included the City Council, enterprise partners, and the city's universities. City Verve was the winner of a £10 million prize from the UK government's joint initiative with Innovate UK (Department for Digital, Culture, Media and Sport 2015) and was part of a wider government investment in the Internet of Things, announced in 2015. The project would see Manchester 'become[ing] world leader in "smart city" technology, to improve life and services in the city. It was set to be a "smart city demonstrator"' (Department for Digital, Culture, Media and Sport 2015: n.p.), a model, a localised experiment that would develop and test various technological solutions offered by the Internet of Things. The other project was called Triangulum, a 30-million-euro project funded by Horizon 2020 EU funding scheme (The University of Manchester n.d.a), a consortium of 22 partners across six European cities, coordinated by the Fraunhofer research institute, and set to work in 2015–2020 (Fraunhofer Institute IAO 2020). Manchester was one of the participating cities in the consortium, taking part in the initial stage as a 'lighthouse' city where smart city ideas would developed to 'demonstrate, duplicate and disseminate' (Fraunhofer Institute

IAO 2020) to other European cities later. The project was led by Manchester City Council and the two Manchester universities (the University of Manchester and Manchester Metropolitan University), in partnership with a lead tech company, Siemens.

Both projects were set to develop their demonstrators in the 'Oxford Road Corridor' (Oxford Road Corridor n.d.a) – the city's 'innovation district' spanning from St Peter's Square where the City Hall and the Central Library are located, through the campuses of the University of Manchester and Manchester Metropolitan University, the Oxford Road Campus of hospitals (Manchester Royal Eye Hospital; University Dental Hospital of Manchester; Manchester Royal Infirmary; Royal Manchester Children's Hospital and Saint Mary's Hospital for Women and Neonates) and Manchester Science Park (Oxford Road Corridor n.d.c). On the Oxford Road Corridor website, which has since changed multiple times, the 'smart city' page had the following statement: 'The Oxford Road Corridor is a test bed for a number of Smart Cities projects that use technology to improve city's transport, health, environment and energy' (Oxford Road Corridor n.d.b). Both Triangulum and City Verve had extensive media coverage at the time of their launch. Both had a very active presence on social media (Facebook and Twitter) and ran a series of public events as well as regular podcasts and other forms of digital dissemination. What were these projects' promises, and what can we learn from them?

### City Verve: the promise of the Internet of Things

Envisioned as a 'platform of platforms' (Royal Academy of Engineering n.d.), encompassing multiple organisations, technologies, and initiatives, all of which used data generated by the Internet of Things, City Verve was structured around four themes: health and social care; energy and environment; transport; and culture, public realm, and community. The project brought together multiple ideas and prototypes and involved many organisations and participants. Among the project's plans were management of chronic respiratory conditions using a sensor network; community wellness via a network of sensors placed in parks and along school routes to track and gamify physical activities; talkative bus stops that combine location-based services, sensors, smartphone apps and digital signatures; smart lighting on bikes to collect data about cycling routes, alleviate congestion, and encourage alternative transport; bike sharing scheme and e-cargo bikes; and monitoring of air quality via street furniture and lamp posts (Jefferies and Anderson 2017; British Telecom 2017).

The project was presented as an example of 'citizen centred' and 'human centred' design that 'puts people at the centre of IoT and the smart city' (Hemment et al 2018: 3) – a somewhat ironic notion given that citizens were only one group of stakeholders, next to central and local government, academia, enterprises, and local businesses (IOT UK 2017). Despite having the theme of energy and environment as part of its focus, environmental concerns were not always explicit in City Verve's presentation – more than anything, City Verve looked like a festival of excited innovation, with various specific initiatives developed locally, and through various partnerships. At the same time, all were united by the idea of an experiment, a trial to pave future ways. Its kick-off launch used the words 'building a blueprint for smarter cities' (YouTube 2016). Echoing the narrative of Manchester as the 'first industrial city', the PR campaign of City Verve used the slogan of 'city of firsts' (Innovative Comms 2017), linking technological innovation to everyday pride of the city's residents. The short film, 'Manchester: City of Firsts', commissioned by City Verve and directed by writer and filmmaker David Petch, links the latest technological innovation to the city's residents' pride in many other 'firsts' – 'the first public library, passenger railway, football league and first life through IVF' (Barlow 2018: n.p.).[3]

Of course, energy and environment were present in some of the initiatives. For example, British Telecom – a company that until recently had sole control of phone and broadband infrastructure in the UK and which was the technical lead in City Verve for travel and transport – stated that the company works with various IoT partners to make travel in the city 'greener, faster and easier' by using sensors at bus stops, and by working with cyclists. Here is one example: a smart cycle lights project which was part of CityVerve:

> With an army of 180 willing volunteer cyclists, SeeSense's ICON smart cycle lights were installed on their bikes and used around the city every day. The cycle lights link up to a smartphone app via Bluetooth and transmit data to the BT CityVerve Data Hub. 'This data collection and sensor communication works in two ways,' explains John,[4] 'firstly, to make the bike light flash brighter and faster in riskier situations, such as crossing busy junctions or approaching roundabouts, and secondly to feed back data about routes taken and the cycling environment'. (British Telecom 2017: 5)

SeeSense is particularly interesting, in showing how the promise of a greener city can be extremely vague. One can say that any initiative that supports cycling is contributing to a greener travel, by reducing road congestion, carbon emissions, and air pollution. At the same time, the focus of SeeSense is on data collection and transmission; and on the real-time 'conversation' between the lights, the smartphone app, and the central data hub. Whether this would lead to an increase in cycling – or, indeed, make a city greener – remains difficult to determine. It is equally difficult to determine whether the project is genuinely driven by environmental concerns, or whether cycling is merely a stage for the excitement of innovation, or perhaps a convenient way to greenwash an investment in developing a new gadget.

It is not possible to fully unravel these motivations without a long-term, and in-depth study of all the actors involved. However, an initial read of SeeSense, as well as many other projects developed under the umbrella of City Verve, suggests that regardless of having the environment as one of its themes, CityVerve is a smart city demonstrator that is technologically rather than environmentally driven. A particular innovation comes first and it is then tested to apply it to specific environmental issues such as traffic congestion or cycling. What's more, while the work of each new piece of tech is described in great detail (what would the sensor, the app, the Bluetooth device, the cyclist, the Data Hub do), the presumed greening effect of the piece is left vague.

In 2018, as the City Verve project came to an end, social media platforms showed celebratory results: connected devices were installed across several buildings; the first talking bus stop was created; smart lights have been deployed for cyclists; innovative powerbank technologies are ready to create 'energy islands'; and much, much more. 'The project is closed but the smart city legacy lives on', says the top of City Verve's Twitter page, no longer updated since 2018, though not yet deactivated or deleted (like the project websites), at the time of writing this chapter in summer 2023. The last several posts on the page, dated July and August 2018, detail the project's achievements in a series of GIF animations in green/blue colours. One image shows the Manchester skyline, with the Wi-Fi icon symbol flashing brightly on top of each building, underneath the City Verve logo and the words '85 connected devices have been installed across 6 buildings' (Twitter/X 2018). Another shows the same skyline, with a moving bus and a bus stop, also flashing a Wi–Fi signal, and the words 'The first talkative bus stop has been installed' (Twitter/X 2018). There are other posts on the same page, specifically addressing energy and the environment: 'One Tesla powerpack will make MSP's bright building an energy island within 12 months'; '85 IOT smart lights have been deployed across the city's cycle network', '211 healthy water sensors, and counting, have been installed'. And there are other posts showing much higher metrics that refer to the data generated: 'Over 1.5 million miles of vehicle data have been collected through our smart vehicle trial', 'Our developer portal has received over 1.8 million API calls to access 200K of project records' (Twitter/X 2018).

The project's branded green-blue colour scheme, coupled with the messages, signal both the promise and the achievement of environmental commitment, even if the scale of environmental interventions is relatively small. But while sensors and other connected devices – the key element of IoT allowing different objects to communicate with each other – are at the heart of City Verve's promise of making Manchester a city of the future, what is strikingly absent is the discussion of the environmental footprint of this innovation itself. For example, what are the energy and infrastructural demands of the data generated? What are the environmental impacts of manufacturing those sensors and well as other objects which are part of the IoT's promise? What would happen to the data and the devices after the two-year project is over?

The lack of attention to this side of the promise is not due to the absence of available data or tools of assessment: for example, one only needs to turn to frameworks such as the 'Life cycle analysis' (Istrate et al 2024; Itten et al 2020) to account for the carbon footprint of a demonstrator project, from start to finish; or engage with the wealth of knowledge about e-waste to form a much more sober approach to the environmental costs of creating multiple, and often short-lived, censors and devices. Accountability for direct and indirect environmental harms would be needed for *both* the longer-term use of the proposed innovations and the short-term smart city experiments themselves. The omission, therefore, is both strategic and systemic – and, as we discuss in the Conclusion, is fundamental to the very nature of smart cities' broken environmental promises.

### *Triangulum: the promise of a sustainable, energy-efficient future*

Contrary to the case of CityVerve, Triangulum had an explicit environmental agenda at its heart. The Triangulum project – a demonstrator focusing primarily on energy efficiency – was set to explore, among other things, 'smart green growth in urban areas', 'zero/low energy districts', and 'sustainable urban mobility' (The University of Manchester n.d.b) using several case studies of smart technologies, such as smart energy management, smart air and traffic monitoring, cycling, and community wellness schemes. On the website of Siemens - the lead technical/industry partner of the Manchester Triangulum team, the project was presented as investigating 'how to balance energy consumption and demand, reduce costs and carbon emissions and increase the use of renewable energy along the city's Oxford Road Corridor' (Siemens 2019). In a more detailed description of the project, published by the Association of Project Management, Triangulum was presented as an answer to Manchester City Council's environmental agenda, and more specifically, its effort to 'find solutions for various causes of inefficiency, waste and pollution' (Association for Project Management n.d.). The same article elaborates on the project strands which are touched upon only briefly on the University of Manchester webpage: 'The ICT element was about connecting everything into a single system. The mobility element included the creation of more charging points for electric vehicles. The energy element involved the implementation of smart and renewable technologies' (Association for Project Management n.d.: n.p.).

The project began with research into energy use to develop a system of smart energy and to establish whether it would be possible to take the area of Oxford Road off the grid, using energy storage and 'smart building controls'. Triangulum consisted of several case studies, where energy-efficient solutions were trialled in different locations in the Oxford Corridor area – a 400kWh lithium-ion battery integrated with solar panels on Manchester Metropolitan University Campus; or a new Building Energy Management System (BEMS) used at the Manchester Art Gallery to control the temperature and humidity vital to the conservation of artefacts and the building itself. A cloud-based energy management platform connected these and several other locations.

In 2019, at the end of the project in Manchester, Siemens published a celebratory description of the project's completion (Siemens 2019). Key members of Triangulum project – academics, art gallery directors, and the mayor – were cited, linking the success of this smart city demonstrator to environmental goals and commitments. The project was named as being key in meeting zero carbon targets by 2038; and hugely beneficial in 'boost[ing] green energy solutions' and moving towards 'a smarter and more sustainable urban energy paradigm'. 'We take climate breakdown very seriously', the deputy director of Manchester Art Gallery was cited in the report, noting the Gallery's commitment to reducing energy consumption and carbon footprint (Siemens 2019: n.p.). Similarly to City Verve, the narrative here is that of a legacy, a future-looking promise. The importance of specific solutions, developed over the course of the project, is presented as lying in their potential to be expanded beyond the trial, and carried on long term. As noted on Siemens' website, 'the pioneering five-year future smart cities initiative … proved [that] low-carbon cost-efficient smart cities are achievable, repeatable and scalable':

The solution optimised energy consumption, reduced carbon dioxide $(CO_2)$ and lessened the area's dependence on the grid. Scaled citywide, the central controller could potentially save Manchester approximately 57,000t $CO_2$ emissions per annum – the same as taking 12,000 cars off the road each year. (Siemens 2019: n.p.)

With the project's future orientation embedded in the very idea of 'demonstrators' and solutions that 'could' achieve things, we must ask, what is their environmental legacy beyond the projects' timeline – in a near future of several years after completion, and in the long durée? How, in other words, can a project move from 'could' to 'will'? To understand that we need to look more closely at the very promise of a legacy; as well as at discontinuities and ruptures that emerge in its shadow.

## The promise of a legacy

When I began working on the Manchester chapter of this book in 2022, I found multiple news stories, as well as subsequent short reports by the University of Manchester, various industry bodies who were involved in the project such as Siemens and British Telecom, just cited, and Manchester City Council itself. All the reports presented brief summaries and referred the reader to project websites for more details. The websites, however, were no longer available. City Verve's URL, http:// www.cityverve.org.uk, led to the domain host site filled with random adverts. Clicking on Triangulum's URL, https://tri angulum-project.eu, showed a blank webpage with a 'there has been a critical error on this website' message. Likewise did the old Manchester City Council website, dedicated specifically to its earlier smart city strategy: following the link www.man chester.gov.uk/smartercity took me to a page carrying the Manchester City Council logo, with the body of the page containing the 'Page not found' message.

The search for detailed and comprehensive official project information for Triangulim and City Verve quickly began to feel like a haunted ethnography (Kuntsman 2007), one that is filled with dead ends, error messages, and other signs of things that once were but are now no longer. In the absence of project websites where one would expect to find all the information about innovation schemes and case studies in a coordinated way, view dissemination details, and access post-project reports and publications, I resorted to finding remnants of Triangulum and City Verve by sifting through snippets of media coverage, blog posts, short YouTube clips, and social media traces on what used to be the projects' Facebook and Twitter feeds. For example, on CityVerve's Facebook page one could see that links had been regularly posted to the project podcast series. The podcast itself was no longer available, but its 'shell' – an image preview and a short commentary on the Facebook post – were still telling part of the story. As I scrolled through the projects' Facebook and Twitter pages all the way back to the time of their launch, I could see a range of exciting promises, some of which are vaguely related to environmental issues, some focused on health and well-being, and yet others more generic and revolved around specific pieces of tech: smart technologies to get people walking outdoors; augmented reality to enjoy art in a university campus; smart cycling with an 'M' scheme – a network of rental bikes that operate with the help of an app – would beat air pollution and reduce emissions.

In 2022, only a few years after the projects ended, the specific details of each of these initiatives, beyond a brief media report or a cache preview of a blog post on Facebook, were very difficult, and often impossible, to find. Most evidence was second hand, in the form of an old media coverage, a photo included in someone else's presentation, or a brief mention in a report.[5] It was hard to see what actually took place or get a proper feel of the two projects. The official 'holders' of City Verve and Triangulum – their websites – were gone; and the semi-official sites of their collective memory in the form of social media sites were patchy and incomplete. Of course, it was possible to recover much of the original content via the WayBack Machine (https://web.arch ive.org/) – the Internet Archive site which stores captures of old webpages taken periodically and available to view even when the original site is gone.[6] However, WayBack Machine offers only a partial answer. The experience of browsing through the archived pages of the two projects was as illuminating as it was frustrating: I was able to recover a lot of the websites' content, but finding further information would often lead to a dead end - a broken link, a missing page, or a partial view.

I frequently felt as if the smart city demonstrators were slipping through my fingers – or were they just someone else's dreams I caught a glimpse of? The two smart city projects at times felt like ghosts (Gordon 1997), invisible and long gone but with their many traces scattered all over the web, leaving an eerie trail in the aftermath of their disappearance. Was I a researcher, a detective, or an archivist when I was piecing the parts together? More than a methodological reflection on this brief digital ethnography, my search yielded questions about the partiality and inaccessibility of the projects' legacy,[7] for the professionals in the field, and for the general public – including the city's residents.

Eventually I was able to contact some of the people who were involved with the projects – a university researcher who led one of the projects, and a senior Manchester City Council employee who was involved in several smart city initiatives – and obtain two detailed evaluation reports which were produced after the projects had ended. They, too, provided only a partial picture.

As I reflect on the elusive nature of the two projects' public vanishing visibility – owing both to the temporality of project lifecycles and to the current infrastructure and platform economy of digital memory itself, where many websites and social media accounts tend to disappear swiftly instead of remaining online for years – I consider the question of legacy as a matter of post-project sustainability. Where do smart city projects begin and end? This is one of the main questions explored throughout this book. One of the key aspects here is what is left after a project ends: what shape is it in, who was it for, and how long did it last? How does one continue to 'duplicate and disseminate' after the initial project is complete? And what are the relations between the project legacies that are envisioned – and that become possible – and the environment?

On the surface, the questions of sustainability and continuity are not ignored. They are clearly addressed in both the City Verve post-project evaluation (Digital Property and Cities n.d.) and the Triangulum 'Smart City Guidance Package', an end-of-project document developed by the international Triangulum consortium (rather than just by the Manchester team, Borsboom-van Beurden et al n.d.). City Verve evaluation focused on challenges to scalability and how it can be achieved, but completely lacked a discussion of environment concerns or the climate change, both when addressing the global, national, and local context of the evaluation (focusing on development, innovation, and the digital industry) *and* when addressing the wider scalability and implementation itself. The discussion of the latter evolves around challenges in resources, governance, infrastructure, and delivery. The Smart City Guidance Package, too, speaks primarily the language of project management and what makes any project (or its scaling up) successful or a failure. The package does note that environmental sustainability is important and that many environmental departments of local governments shift:

> from an approach focussed on immediate environmental quality towards the more holistic and long-term consideration of sustainability: not only now and here, but also elsewhere and later. Energy efficiency, climate change and scarcity of resources have become integrated parts of most sustainability and environmental plans and are therefore closely related to smart city plans. (Borsboom-van Beurden et al n.d.: 30)

And yet the guidance on achieving that is generic, rather than specific. Strikingly, neither of the documents address the overall environmental impact of digital technologies involved in smart city initiatives, the global supply chain that they rely on, the resources required for their operation, or the potential mountains of e-waste they might leave behind.

Both projects position themselves as a taster, a trial exploration that holds a promise of replication and broader benefits – 'achievable, repeatable and scalable' – if we use the language of Siemens report cited earlier. But behind the promise of scalable solutions leading to a greener, climate-responsible future lie two paradoxes. The first one has to do with project life-cycle: the developments, envisioned as future-oriented promises, only last as long as the project funding lasts. The issue of dependency on funding came up clearly in the interviews. For example, my interviewee from the organisation called Open Data Manchester noted that many smart city projects were driven by European research funding, such as the European Commission's 'Information and Communication Technologies Policy Support Programme' and 'Horizon Europe' programmes (interview, March 2003). Another interviewee, who worked for Manchester City Council and was involved in smart city initiatives for over a decade, explained that one of the main struggles for the Council is sourcing sustainable funding for smart city innovations, and moving away from EU research funding schemes which are no longer available to UK institutions since Brexit. Whereas many people working in smart city projects would have liked to have seen more project sustainability, this problem was part of the bigger issue of dependency on external funds. In some respects, as my interviewee explained, it was UK-wide policy failure, inevitable where both private and public investments are insufficient, in particular in cities outside of London (interview, March 2003).

Project-based temporality, however, is not only about continuity of funding and resources available to carry on the developments that were piloted. It is also about barriers to accessing raw project data, prototypes, and models created during a project, *after the project is finished*. For example, data sharing was at the heart of City Verve at the time the project took place: 'Crucially for the CityVerve project, the data gathered [as part of the SeeSense project] is also aggregated and then made available to the wider CityVerve consortium' (City Verve n.d.). And yet, the data was no longer shared with other developers or the City Council later, as mentioned by yet another interviewee, who was a member of the Manchester

City Council Digital Strategy team in 2023 (interview, March 2023). The smart city initiatives, thus, are ghostly not only in their disappearing presence in spaces of institutional and public memory – on the web, on social media, on the Internet archive. Their data, prototypes, and insights are ghostly, too – and, as we will see in the next section, they almost disappear from view in the city's most recent digital strategy.

The second paradox relates to the technology itself. As noted in one piece of coverage of Triangulum: 'One of the biggest challenges for the project was the speed of technological advancement. Over the course of five-year project, it's easy for technology to go out of date, rendering the findings useless' (Association for Project Management n.d.) A smart city project, it seems, holds a built-in temporal contradiction: aimed to create the most innovative and most experimental technology, it operates within a framework where each experiment is always already outdated. Crucially, this is not just about the built-in failure of innovation in a rapidly changing environment. Rather, it is about the global and the local environmental costs of always chasing a new legacy that already goes outdated while still being tested. For example, the relations to resource extraction and e-waste are set as unsustainable and disposable by design (Chen 2016) – since none of the experimental gadgets are designed to last – paradoxically, even in the initiatives that specifically focus on sustainable energy or green transport. The promise of a legacy, thus, is not merely a phantom – it comes with very tangible environmental harms.

## New promises, disjoint priorities

Between the second half of 2020 and early 2021, still in the grip of the COVID-19 pandemic but already considering post-lockdown 'recovery', Manchester City Council circulated a draft of its digital strategy for consultation (Manchester City Council 2021). In 2022, the new 2021–2026 strategy was officially published, presented as 'a new digital vision for Manchester', detailing the city's approach to digitisation (Manchester City Council 2022). In the opening pages, the policy acknowledges that over the last ten years Manchester was involved in European projects 'with a focus on becoming a "'smart city"', and has now moved from a separate smart city strategy to a broader 'digital strategy' (Manchester City Council 2022: 5). The latter seemed to shift in both tone and scale, from the focus on experimental innovation to 'people, places, prosperity and sustainable resilience', where the smart city is one part of a broader vision. Although the vision nods at the previous projects (see Figure 3.2), there is no narrative of continuity here – neither of the City Verve or Triangulum legacies to carry and grow their demonstrators, nor of the lessons learned where a scaling up was particularly promising or has run into a difficulty (as mentioned, for example, in the Triangulum final reports, see Borsboom-van Beurden et al n.d.; or in Fraunhofer Institute 2020). It was as if the last ten years were

**Figure 3.2: Manchester Digital Strategy 2021–2026**

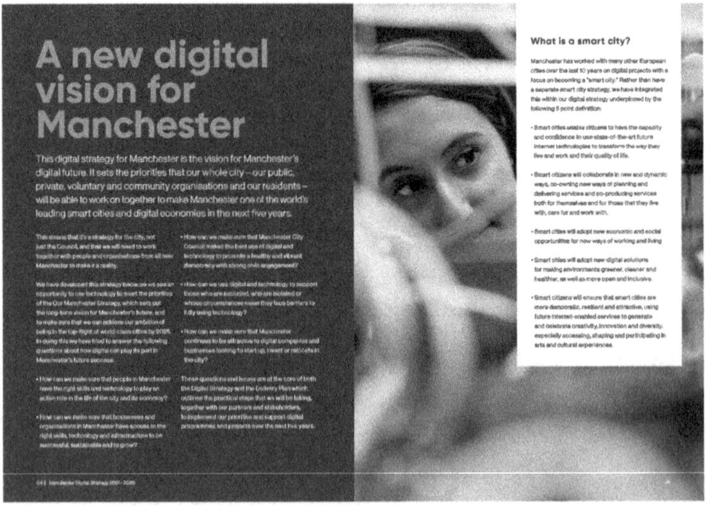

Source: Manchester City Council (2022)

now in the past and no longer relevant; and the new strategy is, literally, starting its vision from scratch. Interestingly, despite its timing, the new strategy is *not* anchored in social, economic, or technological changes brought on by the COVID-19 pandemic, in a striking contrast to smart city initiatives developed in Helsinki (discussed in the previous chapter), and indeed many cities globally, where the pandemic was seen as a catalyst for, and a justification of, accelerated digitisation.

It is also interesting to see the place of environmental and climate concerns in this new strategy – simultaneously explicit and vague. On the one hand, the new strategy reads like a climate-conscious pledge: 'We will achieve are zero-carbon ambition by 2038 at the latest, via green growth, sustainable design, low-carbon energy, retrofitting buildings, green infrastructure, and increasing climate resilience' (Manchester City Council 2022: 8). On the other hand, the elaborations on the 'sustainable resilience' aspect of the strategy use economic and environmental sustainability interchangeably, with the former seeming to be the main drive, and the latter always taking a back seat. Not surprisingly, the contradiction between the two forms of sustainability – that is, that economic growth and sustainable profit are key culprits of environmental degradation and the climate crisis – is not mentioned. Instead, the strategy places greater emphasis on improving access and reducing inequalities. The latter is manifested specifically in addressing the digital divide in the city, although without necessarily elaborating how bridging the divide would work, or acknowledging that digitisation can at times exacerbate, rather than solve, social and economic inequalities, for example when it comes to the growth of digitisation in racialised policing that explicitly harms Black and other communities of colour (Williams 2018); or the rising use of automated online services and algorithmic decision making in the benefits/welfare services, which inflict both emotional and economic harms on those most vulnerable and most dependent on the state (Kuntsman and Miyake 2022).

Although the main declaration of the new strategy is about the commitment to 'become a zero-carbon city by 2038' and 'meet the many changes presented by the climate change' (Manchester City Council 2022: 42), most of the details in the strategy are about the *infrastructural* and *commercial* resilience of technology and connectivity. The necessity of digitisation itself is never questioned; instead, the vision presented is to 'adopt and adapt to' new and rapidly developing technologies; and to provide growing and inclusive connectivity and access to digital resources for all citizens, and to ensure that the city itself is 'future proof' to be fully digital. The environmental commitment itself is unclear, beyond declarations of zero carbon targets and the overall 'climate resilience'. At the same time, digital technologies are presented as key solutions to the environmental targets:

> Digital technologies have a specific role in supporting action on climate change and zero carbon targets. Digital can be transformational for the environmental agenda and offer very practical solutions for current and future action. New, low carbon opportunities can be realised through enhanced digital connectivity and data analytics, especially in areas such as mobility, logistics, food and buildings, supporting sourcing of energy from green sources to improving air quality and encouraging more walking and cycling. Digital can also help create a new smart circular economy where local sourcing is the norm and product information can be made more accessible and easier to analyse with repair and recycling facilities more widely known and understood, with makerspaces and other digital production facilities able give products longer lifecycles. (Manchester City Council 2022: 42–43)

Just as in the previous instances of smart city demonstrators, the new digital strategy does not consider the environmental tolls of digitisation itself, and that despite the fact that among the body of evidence listed as informing the strategy such as various reports and policy papers, is *Smart and green: Joining up digital and environmental priorities*, a report by the Green Alliance (2020). *Smart and Green ...* is so far the only UK policy document that openly and extensively addresses both the digital *and* the environmental agendas in the country's post-pandemic recovery, including an acknowledgement of the current disconnect between the two. The document clearly states that 'digitisation policy does not prioritise environmental goals' (2020: 8); and that digital technologies themselves 'are currently linked to substantial climate and resource impacts' (2020: 9).

It is these impacts that were at the heart of my conversations with several members of Manchester-based organisations who worked on either digitisation or climate/environmental issues (tellingly, there were no organisations that simultaneously tackled both). One of the interviewees, the CEO of Open Data Manchester, who was involved in various smart city initiatives, and is currently running several projects that support citizen's understanding of, and access to, data, was sceptical about the promise of digital technologies to offer environmental solutions. Smart city initiatives, he explained, are 'often driven by specific technologies and predatory business models', [they] may be offering solutions to something, but not actually tackling the problems that the city needs addressing. And the techno-cultural drive to always seek 'bigger and faster' solutions is in itself not environmentally sustainable (interview, March 2023).

Looking at the digital versus the sustainability agendas more specifically, another interviewee, who has worked at Manchester City Council and was involved with smart city initiatives over the years, noted that although the IT team does have a sustainability agenda, this is not always at the top of the list, with the main difficulty being that 'we don't always have clear sight of what the environmental costs as well as the environmental benefits are' (interview, March 2023). An absence of proper comprehension or assessment models of costs and benefits is apparent, as he details his understanding of the impact of digitisation, whether it is the carbon and heat footprint of data centres, the e-waste generated by smartphones and computers, the energy-hungry and carbon-intensive web design, or the supply chain involved in equipment and consumables, used by the Council team. All these are yet to make their way to a clear and detailed agenda, he explained in the interview, because:

> people on the digital computing tech side often have good intentions and they would like to do more environmentally, but they don't really know what it all means or how to do it. And … people on the environmental side …, can sometimes be technological Luddites. (Interview, March 2023)

The disconnect between the digital and the environmental priorities showed up in other interviews as well. One of the interviewees, from a Manchester-based organisation that has a detailed and extensive agenda on climate change, with a particular emphasis on decarbonisation, noted that there was little to no attention to the role of digitisation in climate change agenda – be it positive or negative. The main benefit of the digital, the interviewee explained, was in the actual technological innovation such as electric cars with their expected impact on lowering air pollution and emission rates, on the one hand, and communication technologies' role in supporting behavioural changes, on the other. For example, if an app can provide live and reliable information on public transport, the assumption is that it would encourage a shift away from driving cars. By contrast, when I asked about any possible environmental harms, these were described as either unknown, or negligible, compared to the benefits the technology can bring. The example provided was that adding 'one more app' to the phone that was already running hundreds of other apps would not make much difference to any carbon footprint. Dedicated to addressing climate change, my interviewee spoke with a strong sense of urgency about the need to tackle major problems such as reducing carbon emissions and achieving net zero, and to adapt to the changing realities of the climate crisis which is currently unfolding. Given the severity of carbon emissions, working out 'a footprint of an app' would be almost laughable, the interviewee explained, whereas the benefit of a 'whole load of new black boxes on streets to monitor traffic' and leading to changes in driving and commuting behaviour seemed apparent (interview, March 2023). With carbon emissions in Manchester – and in the UK as a whole – remaining very high and their reduction not yet on target,[8] my questions about the footprint of tech – framed here as a footprint of 'just one app' – appeared to be insignificant, and almost distracting. The interviewee acknowledged that the broader impact of digitisation and the footprint of the digital economy should be the responsibility of the tech companies.

The digital seemed both vague and not particularly relevant to the urgent matters of climate change, that the interviewee spoke about with care and passion. In shifting away our conversation from the environmental costs of digital technologies, the interviewee emphasised that we must not merely combat or slow down climate change, but, instead, actively adapt to the new reality that is coming in the near future, such as floods or other changes in weather which affect not only quality of life but the global supply chain and can quickly lead to shortages. The important question should address the role of technology in helping us become resilient in view of these changes.

We never got into much detail about digital harms. I felt as if I was speaking a different language to that of my interviewee, despite both of us caring deeply about climate change and the environment. In the preliminary list of questions to be discussed, sent to all the people who I interviewed ahead of time, the footprint of digitisation was broken down in detail.[9]

And yet, the conversation kept moving away, reminding me once again that 'thinking like a climate', to use Hannah Knox's words (2020), often omits thinking about the digital as a culprit, be it the carbon emissions associated with digitisation or any other environmental harms.

While the interviewee from Manchester City Council explained the disconnect between the digital and the environmental agendas with the lack of sufficient and deep knowledge on either the digital or the environmental decision makers, the interview just discussed re-emphasised for me the power of a strong and persistent belief in technology's environmental and climate neutrality. This is a belief that digital technology, while instrumental in addressing and mitigating environmental harms, has little or no environmental footprint of its own – a belief endemic to the field of environmental sustainability (Kuntsman and Rattle 2019) and broader social perceptions of digital technologies. Challenges to this belief were at the centre of my conversation with another interviewee, who was involved in climate change work with the community and the student union in Manchester and the region for many years; including a local charity and a regional organisation both dedicated to improving access to natural environments for local communities. Echoing other conversations, he noted the disconnect between environmental and green agendas and commitments of different organisations, all the way from local organisations and groups to how smart city initiatives are funded and governed. He also called for proactive conversations about smart cities that should involve groups within governmental bodies such as DEFRA (UK Government's Department of Environment, Food and Rural Affairs) 'rather than leaving it to the Department of Culture, Media and Sport' which is currently in charge of all digital initiatives and innovations (interview, March 2023).

Speaking about his current role which involves supporting communities' access to nature, the last interviewee brought to the table an issue that is strikingly absent from most other policy documents and digital visions: the need for physical spaces to accommodate the growing data centres and digital infrastructures, and the impact that taking space would have on disadvantaged communities. In addition to the general question of power demands and carbon footprint of data centres, the question of space for him was in direct relation to shrinking green spaces in the city, which infringes on wildlife and reduces biodiversity, on the one hand, and increases inequality in the physical and mental well-being of different communities, many of whom already have a very limited access to green spaces. The socio-environmental inequality – both local and global – as he explained, would be exacerbated if the already limited access to nature and green spaces would decrease even more; and would be further deepened as a result of e-waste generated by growing smart cities, as e-waste would be dumped on poorer communities in Manchester and the region, as well as on countries in the Global South. This concern resonates deeply with Ushnish Sengupta and Ulysses Sengupta's observation that smart cities do not simply increase the amount of e-waste generated due to having a growing population with a growing amount of electronics, but shift e-waste generation from individual to institutional (Sengupta and Sengupta in progress) – that is, e-waste generated by the large-scale, citywide adoption of digitally connected devices that would turn into what many coin 'the internet of trash' (Higginbotham 2018) – but this issue is rarely addressed in smart city visions. Furthermore, and crucially, my last interviewee's observation reminds us that any form of digitisation and technological innovation is embedded in local and global social injustices (Sengupta and Sengupta 2022a).

## Conclusion

I began this chapter with several key questions that guide this entire book: what are the environmental promises of Manchester as smart city? Where and when do these promises begin and end? What happens when these promises break or are made in such a way that they become violent? And what is left in their aftermath?

Throughout the chapter, I have focused on several promises, and looked at how they were narrated and where these narratives could be found. In my analysis, I was first and foremost driven by the question of the environment. That, in itself, required unpacking: what exactly is meant by 'environment' in each project, vision, or narrative; how does 'environment' relate to climate or nature; and what is the place of environmental concerns in each of the projects? As this chapter has demonstrated, just as the case of Helsinki (discussed in Chapter Two), most smart city agendas in Manchester were driven by specific technologies presented as new and exciting 'solutions', rather than by an environmental need, or by environmentally driven priorities. For example, City Verve focused on the Internet of Things and what it can offer rather than on the environmental problems and solutions they might need. Furthermore, the environmental agendas themselves were segmented, even when their promise was generic; for example, Manchester's latest digital strategy 2021–2026 used the language of 'climate resilience' and 'climate change', yet focused almost exclusively on decarbonisaton or green energy.

A related issue, emerging here, concerns the relations between broader concepts such as 'environment', 'sustainability', or 'climate change' and their specific manifestations: for example, biodiversity was not mentioned at all in all the materials collected except for one interview; e-waste only mentioned in passing a few times but not embedded as an issue in any of the demonstrator projects; heat emissions of data centres were not discussed at all except for one interview, where such emissions were presented as a potentially positive thing. In the interview, a member of the Manchester Digital Strategy team noted that the heat generated by data centres is currently 'wasted' and needs to be used for heating houses, thus suggesting that heat emissions of data centres are a resource rather than a form of potential harm (interview, March 2023).[10] Similarly, 'sustainability' – which is perhaps the main buzzword in current discourses around urban planning and environmental and climate concerns – was mentioned repeatedly in a number of policy documents. However, it was mostly referring to infrastructural, commercial, and economic sustainability – that is, ways in which the infrastructure needs to grow to support the increasing 'need' for digital connectivity; or ways in which smart cities would allow sufficient and ongoing economic gain rather than an environmental one.

I conclude the chapter by returning to the question of what is left in the aftermath of the smart cities' promises. The answer to this question has been on my mind, as I cycled daily through Manchester, where the contrast between shiny smart enclaves of the city centre and the deprivation of some central and many suburban areas is striking; where traffic and air pollution have long reemerged from the short-lived drop seen during the COVID-19 lockdowns, and are at an all-time high; where poverty and homelessness are constantly on the rise, and so is digitisation, jittery yet persistent, all-encompassing, but not necessarily smart, or smart but not at all kind or caring. I am thinking about the smart cities' aftermath, as I arrive to the university campus on a bike, passing the more well-off southern suburbs, and the more deprived areas located on approach to the Oxford Road Corridor. On my way, I navigate multiple potholes, typical of many Manchester roads, pavements, and cycle lanes; battle the traffic, and inhale enough car exhaust fumes to ensure a sore throat for the rest of the morning. The last five-minute leg of my journey is very pleasant, though, as I pass through the Oxford Road Corridor – a space that is 'regularly maintained, signposted, paved and landscaped to facilitate movement, but also to present a slick urban aesthetic', as Emma Fraser and Clancy Willmott note (2020: 357) when they discuss the relations between urban innovation, wasteland, ruination, and nature.

On Oxford Road, both the road and the cycle lane are now much wider, the buildings are tall and shiny, and the pavement clean. I am greeted by a digital display (Figure 3.3) showing the daily temperature and a count of how many bikes have passed the road on that day. The data does not seem to be doing much apart from click-counting – in that respect, it is another example of what Liu Xin has described as 'passive data' in the previous chapter. And yet, the display itself seems to act as a boundary object, a reminder that one is entering a 'smart' zone where steps and wheels are documented, and where digital technologies are served to tell a picture story. Of what, not entirely sure .... For those on the privileged side of Manchester's many divides – across lines of race, religion, or poverty – the display is a gate marking the entrance into the 'innovation corridor' and/or the university area, a smart city oasis where several space- and time-contained legacies might still be present, perhaps inside a university lab or a gallery. For others it is likely invisible; or perhaps it marks a zone of unattainable wealth.

I wonder who gets to benefit from these legacies, where and how. Where does one find that promised smart city which should be here by now – one with clean(er) air, hyperconnected parks, renewable and cheap energy, and talkative bus stops, designed during the now completed 'demonstrator' projects of the previous decade? Unlike many other cities, most of Manchester's bus stops do not even show real-time updates or arrival times, and the buses themselves do not have a computer screen inside showing stops on route, or voice announcements. Manchester still has one of the highest rates of air pollution in the UK. It is still trying to *become* smart, but how? Another new strategy, another hope, another promise. Some of these promises float in the air, unable to materialise without external funding or sustainable governmental or corporate support.

Others are already haunted by the ghosts of their predecessors' failure. For example, when a new smartphone-operated bike rental scheme, 'Bee Network Bike', was launched in 2021, the city Mayor Andy Burnham urged residents 'to not chuck hire bikes in canals' (Alexander 2021). His plea reminded the

**Figure 3.3: Bike counter, Oxford Road Corridor**

residents of the first scheme, 'Mobike', launched in 2016, in the midst of the 'smart city' hype, and widely celebrated. The scheme consisted of multiple docking stations spread around the city centre, offering bikes to rent via a smartphone app. It was promptly discontinued after just one year, when the city had to pull out of the scheme due to many instances of 'vandalism' and 'crime' – a skilful euphemism to both describe and hide Manchester's polarised population – the flashy, digital centre and the deprived and struggling neighbourhoods and towns of the Greater Manchester boroughs. Parts of Mobikes since then were seen rotting in the city centre canal (Coyle 2018), and some of the bikes were ridden, hacked into, and de-networked in some of the city's suburbs, where deprivation and poverty are high and the main forms of digitisation and smartification are overt and covert over-policing with CCTV, biometric databases for stop-and-search, and social media surveillance (Williams 2018). The striking socio-geographic injustice of smart city projects is rarely discussed when talking about digital innovation. Similarly absent from the public conversation is the environmental toll of smart city waste, created due to planned obsolescence or project failure: the metal waste of broken Mobikes or the e-waste generated by the many sensors and digital devices, deployed during the projects' lifecycles, which are now finished, and ghosted from existence.

I wonder if our discussion of smart cities' environmental promises should include not only the new and the innovative, but also the old, the disappearing, the broken, and the decaying. As Fraser and Willmott (2020) have powerfully noted, Manchester Smart City is perhaps best viewed through the lens of the city's many ruins. For Fraser and Willmott, urban ruination and green growth, existing side by side with new shiny technologies and buildings, are examples of the visual contrast between 'the push for smartness and the messiness of everyday life' (2020: 362); and sites where disfigured Mobikes are left to rot are not just vandalism but a 'playful rejection of the smart city' (2020: 362). Their visual and poetic account of urban ruins in sites of smart city innovation is a refreshingly critical intervention into the celebratory industry and policy narratives of smart city innovation. However, I would like to push their intervention in a somewhat different direction, away from the challenge of smartification narratives through the 'messy' and the 'playful'. Instead, I propose that we use haunting (Gordon 1997) – signs of absence, discontinuation, erasure, and forgetting – as well as forms of material, *physical* brokenness, such as ruins, bike parts, and e-waste, as an entry point into the broader conversation about failure, brokenness, their violence, and their affective and political power. These are questions we turn to in the final chapter of the book.

# Conclusion: In the Ruins of Broken Promises

## Introduction

This book began with questions about smart cities, digitisation, and the promise of digital technologies to support the environment and ensure a brighter future for all, despite the harms these technologies might inflict. Inspired by Berlant's (2010, 2011) notion of cruel optimism – an attachment to a fantasy, dream, or promise that is not only unattainable, but can be explicitly damaging or toxic – we traced optimistic narratives and visions of smart cities, and of digitisation, in academic literature and in two cities, Helsinki and Manchester. In Chapter One, we looked at the ideas of, and investments in the story of smart urbanism as a story of progress, and of digital technologies as game-changing tools of environmental sustainability. To challenge these ideas, we briefly explored the untold and invisible stories of environmental harms and social injustices in the processes of resource extraction or manufacturing, which are necessary for smart cities to operate.

We then turned to our two case studies. In Chapter Two, Liu Xin explored the Smart Kalasatama district of Helsinki, drawing on a range of materials including autoethnographic accounts, reports, and audio-visual materials from Kalasatama webpages. The chapter focused on the environmental aspects of smart city, asking what kind of human-digital-environment relations are at work in the imagination, narratives, and practices of making Kalasatama 'smart'. The chapter made visible the relations – and tensions – between the multiple temporalities of Kalasatama: the promise of one more hour a day, the temporality of long-term sustainable development, the short duration of project economy, and the multiple temporal expressions of the digital itself. The chapter traced the social and environmental temporalities that take place outside the demarcated boundaries of smart cities, attending to the schism between the promise of speed and real-timeness of digitisation and the less visible, and often destructive processes such as digital ruination, the sedimented and slowly unfolding environmental changes, and the long-term impact of energy transition.

In Chapter Three, Adi explored policy narratives, research documents, and media discussions of several smart city initiatives in Manchester which have recently ended, as well as interviews about the latest developments around smart city initiatives. The chapter traced the promises made about technology, innovation, and the environment, and focused on the disconnect between the digital and the environmental concerns. This disconnect, as the chapter demonstrated, occurred on the level of agenda setting and policy, as well as in the complex reality of actual smart city initiatives and their after-life. The chapter was particularly attentive to the silences around environmental harms and the invisibility of local and global injustices which are exacerbated by digitisation. It discussed the lack of continuity between various smart city projects that began with promises of replication, scalability, and long-term legacy, but turned into fragmented pockets of knowledge and ghostly digital remains. Echoing Liu Xin's discussion of digital ruination as a deeply felt form of brokenness, emerging as incommensurability between the speed of digital production and its disposability, Adi's chapter raised the pressing need to consider the material brokenness and the ruins – electronic and otherwise – left by the smart city projects.

## The cruelty of failure

In their *Cruel Optimism* (2011) Berlant argued that 'an optimistic attachment is cruel when the object/scene of desire is itself an obstacle to fulfilling the very wants that bring people to it: but its life organizing status can trump interfering with the damage it provokes' (2011: 227). Throughout the book, they explored *how* and *why* this attachment persists on both the individual and the collective/national level – be it out of habit, hope, or 'investment in the possibility of repair' (2011: 227). Berlant noted that seeking repair 'of what may be constitutively broken' is an exhausting, politically depressed position (2011: 227), and yet, its hold is exceptionally powerful. It was this depressing, exhausting but powerful investment that we have engaged with in this book. Throughout the book, we were particularly attuned to the complex but often elusive relations between the affective power of smart cities' environmental promises – the insistent belief and investment in the promise of digital technologies, despite existing critique and awareness of its limitations – and the political, economic, and environmental cruelty of this optimism.

To understand how cruel optimism works in the case of smart cities' environmental promises, we need to investigate how the idea of 'failure' is normalised in smart city projects, as either invisible or inevitable. Sometimes, smart city initiatives promote an idea which had already failed in the past. The repetition of what had once failed – the 'let's try again' – becomes a powerful force to both cement and ignore the reasons for previous failures: the failure is simultaneously acknowledged and erased. In Berlant's words, this is an optimistic attachment par excellence, involving 'a sustaining inclination to return to the scene of fantasy that enables you to expect that *this* time, nearness to *this* thing will help you or a world to become different in just the right way' (2011: 2, emphasis in the original). Except smart city projects are not merely a fantasy. Drawing on significant funding – whether governmental, municipal, or corporate – they create powerful social-material realities. Such was the story of smart bikes, described in Chapter Three: a new initiative for smartphone-operated bikes was launched in Manchester shortly after the previous one had failed. When the same service was reintroduced a few years later, the city mayor was repeatedly cited in the media, urging the residents not to let the new scheme fail, too. Manchester had pulled out of the first scheme allegedly due to the high level of 'vandalism', where many bikes were hacked into, stolen, or broken and left to rot in canals or fields. Such a response by city residents has been theorised as a 'playful rejection of the smart city, an act that leaves the bike covered in muck – mired in decomposed, composted urban filth, the residue of centuries of innovation gathered in the sludge at the bottom of the canal' (Fraser and Willmott 2020: 360).

Breaking and returning a piece of networked technology to nature, albeit in a polluting and toxic way, can be seen as an opposition to the sleek, high-tech image of travel in a smart city. However, we propose to also consider it as a form of resistance to what the bikes – and smart cities' initiatives more broadly – stand to represent: long-standing and growing economic inequalities, poverty and deprivation that shape many of Manchester residential neighbourhoods, which stand in a striking contrast to the over-investment in the city centre with its businesses, tech companies, hospitals, and universities. Breaking into, stealing, and 'vandalising' a smart bike is first and foremost a cry out against the uneven distribution of resources and the cruelty of racialised and classed priorities of what, who, and where should be 'smart'. Seeing the failure of the rental bike scheme not as vandalism, or playfulness, but as an act embedded in systemic injustice, is not about justifying or condemning it. Rather, it sharpens our view of the power – and the citizens' experience – of repeated investments, despite the earlier failure. Hidden behind the seemingly naive optimism of hoping that what did not work a few years ago, would work now, is the cruel geography of digitisation and innovation, that invest in *some* residents, *some* needs, *some* priorities, without attempting to the others.

Other times, smart city failures are rendered not only inevitable but desirable, as demonstrated in Chapter Two, when looking at Kalasatama's notions of smart city as experimentation. Experimental environment, as we learn from Kalasatama, not only allows for, but in fact requires failure. Following the typical conception of scientific experiments, trial-and-error is seen as the essence of innovation, as it testifies to its value and credibility. It is perhaps not surprising that many funding schemes for innovation projects ask in what way the projects include high risk factors. As Sianne Ngai (2020) notes, the risk of failing intensifies the gimmick and the spectacle effect of experiment. In the case of Kalasatama, the small-scale and short-term pilot projects were justified because their trial-and-error was considered necessary for long-term innovative success – and so was the vagueness of their promise. Tellingly, the Kalasatama smart district report defines the experimental approach to green infrastructure as follows: 'Kalasatama has served as a test area for co-development and smart solutions. Kalasatama has been the first district in Finland to test the regional green factor in urban planning. It calculates and guides the quantity and quality of green spaces in the planning of an entire district' (Forum Virium Helsinki 2021: 18). Liu Xin examined the augmented reality (AR) app developed by 'Green Kalasatama' project in which residents could visualise and imagine how the green spaces in Kalasatama could look like. As Liu Xin suggested, the experiment could be seen as designed to fail because what happened after the collection of data was intentionally left unclear. The project justified this as a necessary component of innovation. It was posited that there would always be more to learn with the development of technologies and the advancement of users' smart devices. Beyond the failure of one particular scheme, we also see the failure of scaling up, often implied in the design of localised, small-scale, and pioneer experiments. Such was the case of many of the initiatives developed as part of City Verve project in Manchester, which was envisioned as a network of small projects using Internet of Things. Many of these had an explicit environmental promise – a talkative bus stop to encourage the use of public transport, a bike sensor to support cycling, smart street furniture to monitor air pollution. None were scaled up, and indeed, all were designed merely as demonstrators of what IoT *could* do. Similarly, the Kalasatama smart district report does not provide any information about whether the green factor – a zoning tool for calculating green spaces in urban planning – first tested in Kalasatama, has been – or would be – further developed. It does not need to. The experimental status of the environment in Kalasatama smart city district explains the scale and duration. Its value lies not in how long the project lasted or whether it was successful, but that it '*serves as a laboratory for new solutions to be tested in Helsinki*' (Forum Virium Helsinki 2021, emphasis by author).[1]

The spatio-temporal configurations of experimentation that we detailed in Chapters Two and Three – short duration, discontinuous but repeated, fast, and abrupt – present a new form of cruelty, one that is not only about the pre-existing socio-economic injustices of urban life but also about the market-driven economy of digital innovation itself. As some have argued, the framing of smart city projects as experiments 'might also hide an agenda which is actually not about sustainable development at all' (Cugurullo 2021: 68). Indeed, the main aim of experimentation is to provide platforms for market creation experiments through funding pilot projects, creating entrepreneurial networks, marketing the city or district, and making it faster to implement the projects by negotiating with existing bureaucratic structures. Most importantly for our discussion here, the experimental logic of smart cities substitutes environmental sustainability and environmental care with an entirely new vision of nature and the environment, that are in themselves either *digital* and *experimental*, or completely invisible.

In Kalasatama district, discussed in Chapter Two, this happened in the following ways: first, the environment was rendered part of a constructed, controlled scenario with limited space, variables, protocols, and predefined problems; and second, it was constituted as digitally mediated and immersive.

For example, in seeing how the green spaces might look like by using an AR application, the residents in Kalasatama experienced the environment through the lens of a smart city experiment. Resembling many examples of employing AR technologies to interact with, learn about, and otherwise 'connect' to nature, the project did not merely digitise the experience of nature. Rather, it offered to experience the promise of technological innovation *before and without* its realisation. As the Kalasatama smart city final report writes, 'As the results of the innovations will not be visible until years later, a new operating model was developed in cooperation between the Smart Kalasatama and Virtual Verdure project' (Forum Virium Helsinki 2021: 18). Crucially, such immersion within a controlled environment also conditioned an amnesia where environmental problems could be rendered out of sight and out of mind, since they existed outside the imaginary world of an experiment. Failure of an experiment, here, was completely disconnected with the harms that might have been inflicted in the process of innovation, and in the aftermath of the project's conclusion.

In Manchester, similarly, one could find IoT experiments where nature would be rendered through the lens of an augmented reality app, or data produced by a sensor, whether it was a sensor placed on a bike light, in the street or in the park. However, more often the environment would disappear completely as the experiments would be confined to specific projects, agendas, and funding investments. The vagueness of what exactly is seen as environmental benefits, and the changing language – 'green', 'resilient', 'carbon neutral' – contributed to this disappearance. The environment might then re-emerge either as abstract data which is shared and managed (Royal Academy of Engineering n.d.); or as a subject of citizen 'learning and awareness of environmental issues in the city' (Fraunhofer Institute IAO 2020: 229). In all of this, long-term success was not expected as it lay beyond the scope of each project. Nor was there any room to enquire about the life-cycle of data or hardware during, and especially beyond, the innovation experiments.

Helsinki and Manchester are not unique. In many ways, smart cities' promises, and the critique of their failures have almost always gone hand in hand. While sometimes hailed for fostering economic, social, and environmental sustainability, the vision of smart cities is often criticised precisely for its all-encompassing vagueness that often ends up being an empty and broken promise. Smart cities as a model for urban development were already widely debated in the early 2000s; and already then, the notion of their failure was at the heart of how smart cities were understood. In the influential article from 2008 titled 'Will the real smart city please stand up? Intelligent, progressive or entrepreneurial?', Robert G. Hollands argues that the 'definitional impreciseness, numerous unspoken assumptions and a rather self-congratulatory tendency (what city does not want to be smart or intelligent?)' (2008: 304) *allow the rhetoric of smart city to promise so much yet deliver so little.* Especially relevant for our book is Hollands' observation of the conflict between the promise of environment sustainability and economic growth. Noting that much smart city's projects' environment sustainability initiatives are often marketing strategies, Hollands asks 'can cities accord the same priority to all aspects of the smart city agenda, or do some elements automatically take precedence over others? (that is, business needs over environmental ones)' (2008: 313). The answer, implied in this question, is that of course they cannot.

Holding the expectation of failure at the very heart of a promise is precisely the paradox of cruel optimism, and it is therefore particularly important to notice how this paradox is normalised. A conference participant once remarked, after our presentation on the broken promises of smart cities, that bureaucratic promises of city planning are often expected to break, and so what? The trivialisation of failure is an important moment, one that we have dwelt on in this book. For despite the expectation of failure, the idea of a 'smart city' continues to attract substantial economic investment from major research funding bodies and, more importantly, from governments and tech giants. Equally, they continue to dominate both the policy discourses and the public imagination. The practices of smart cities are conditioned upon and produce specific material arrangements that have profound environmental implications that might not be immediately perceivable, both in terms of when and where the effect take place, and who and what are impacted.

The real question, then, is not 'so what?'. This is not to dismiss the importance of work that examines how smart city projects fail and what strategies and practices could be implemented to make such projects work. Take for example Ushnish Sengupta and Ulysses Sengupta's (2022a) incredibly important examination of the government-supported smart city initiatives such as Amazon HQ2 project in New York and Sidewalk labs quayside project in Toronto. For Sengupta and Sengupta, the failure of these smart city initiatives lies in institutional practices, including unequal distribution of risk and benefits, and conflicting values between the government and citizens. Sengupta and Sengupta's work is insightful in understanding the schism between smart cities' promises and their lived reality, including injustices inflicted on communities who live in those cities. Yet, their analysis rests on an assumption of direct casual relations where failures are external to the promises, that is that they can be improved or even fixed, with a different design approach. The latter would be based on more/better data, funding, management, and technology – and, crucially, on a community-oriented approach to risks and benefits of smart cities. However, the focus on design, or on community benefit agreements, does not question the cruel techno-optimistic framing itself – the framing of success and failure that informs smart city's narratives, practices, and imaginaries. Nor does it address why and how the idea of smart city sustains its affective and political hold, even as it is vague, empty and at times, undeliverable.

To better understand the complexity of this hold, we turn, once again, to the notion of broken promises. The theoretical framework of broken promises that we developed in this book eschews the opposition between success and failure and the techno-optimism it forms. Instead of simply asking how smart cities fail or what can make them fail less, we historicise and contextualise smart promises in specific socio-economic, technological, and environmental settings, and ask, specifically, what breaks when cities try to become smart. In doing so, we call to reconfigure the debates on smart cities and their narratives of technological promises, and to un-anchor the foundation of cruel optimism by reorienting the storytelling of digital urbanism away from tech giants, experts, and governments, and towards the environment, such as ecosystems, land, air, minerals, and people. By attending to the question, *what breaks when cities strive to become smart*, we argue that it is imperative to simultaneously consider both symbolic and material forms of brokenness, but also ask, what *remains* after the promises are broken.

## What breaks?

### *Broken geographies*

As we look back at the case studies of Helsinki and Manchester, we note that in many visions a smart city is branded as a detachable model, a product that can be exported and installed anywhere. Such is the nature of 'experiments' and 'demonstrators', that appear to be scalable and replicable, irrespective of material and environmental conditions of specific contexts and locations. In reality, as both our case studies show, smart city projects do not just differ from each other, but are in themselves *internally spatially fragmented* – a phenomenon we call broken geographies. In Helsinki, for example, both Jäätkäsaari and Kalasatama districts are smart city pilots, but with different orientations.

Whereas Jäätkäsaari is a test site for smart mobility initiatives that aim to improve the traffic system, Kalasatama focuses on green and sustainable technologies and practices. Internally, Kalasatama's practices for transforming energy infrastructures are also disjointed and fragmented, involving different spaces, scales, stakeholders, and funding schemes. For example, the Internet of Things trials are part of the EU Horizon 2020 funded project, similarly to Triangulum in Manchester. The home automation system in Kalasatama is installed by electrical contractors, who are separate from the Horizon project, and are not responsible for the implementation of energy-related applications and services once the residents have moved in. And the transformation of energy generation infrastructures such as the heat storage facility on Mustikkamaa, a nearby island connected to Kalasatama by the Isoisänsilta Bridge, is carried out by the energy company Helen, owned by the city of Helsinki. The lack of coherent and integrated planning of the home automation system results in its low usage and the ineffectiveness of the energy-saving measures. The broken geographies can also be *felt* as a dissonance. For example, there is a stark contrast between the visualisation of green spaces in virtual, digital forms (such as in the AR app, Green Kalasatama, discussed in Chapter Two), and the actual lack of vegetation in the physical space. This dissonance begs the question of how the environment is understood and experienced by different actors, and what its role is – as a physical space and not just a digital manifestation – in smart city planning and development.

In Manchester, the sense of broken geography was even more apparent. The two smart city projects, discussed in Chapter Three, were both situated in the 'innovation district' area of the city – the area called the Oxford Road Corridor. Located on both sides of Oxford Road, it spans between the city centre, the Oxford Road Campus of the National Health Services (NHS) hospitals and Manchester's two universities. As Fraser and Willmott note, the exact mapping of the area is somewhat confusing: 'In some sources, the district expands on an existing area of health and science innovation adjacent to Oxford Road; in others, "Corridor Manchester" is a wider commercial zone dedicated to innovation and enterprise' (2020: 354). The contradiction and overlap between public services – education and healthcare – and private enterprise and business are worth noting here, for they point precisely to the conflation, so often present in smart city narratives, between public good and private gain. This conflation, of course, is not unique to smart cities – as José van Dijck, Thomas Poell and Martijn de Waal (2018) have argued in their book, *The Platform Society*, the tension between public values and private interests lies at the heart of platformisation of economic, political, and social life – and, as we would add, of today's digitisation more broadly. In addition, specifically to the UK, the conflation of public services and business interests – or their collapse into each other – is part of the broader ongoing process of the gradual privatisation of public education and healthcare systems.

Broken geographies, here, are not just about contested ownership and interests of the area where Manchester smart city initiatives were developed and tested. Rather, we need to look at the geography of prototyping itself. Since academics from both universities were involved in the projects, one might argue that using university campuses and buildings to develop smart city experiments was expected, and convenient. But in doing so, the projects naturalised smart city innovation as placed within academic *and* business environments, rather than in residential neighbourhoods, homeless camps, or food- and clothing-market areas – all of which are located in the immediate to moderate proximity to the 'innovation district' but economically and culturally are worlds apart. In that respect, the geography of smart experimentation has intensified, rather than mended, the already broken geography of the city, where wealth and poverty, privilege and marginalisation already exist in proximity and along invisible lines of segregation.

Given the brokenness of smart cities' geography, the question of 'scaling up' and replicability requires careful revisiting.

In addition to asking whether scaling-up, duplication, and dissemination of various smart city models is possible – or whether they were designed to fail – we should also ask, what are the consequences for social justice and community *and* environmental well-being, when a scaling-up process starts with the university and business innovation park as its model to duplicate elsewhere? Sengupta and Sengupta (2022a) point to the core of such built-in brokenness in their discussion of how government-supported smart city initiatives fail when they do not take into account the unequal distribution of risks and benefits to the local communities. And, as we shall add, we cannot consider environmental justice of digitisation, if it is blind to the geography of social injustice into which the digital is parachuted. Such geography can be local, for example, when we look at the location of data centres, which are constantly expanded, to support the growing digital infrastructures of smartification. In Manchester, the current plan is to build more data centres in the suburbs of the city (interview, March 2023; see also Chapter Three). What might be seen as a potentially positive move – a shift of smartification from the well-off, business, knowledge, and enterprise-focused city centre outwards to the periphery and the suburbs – requires critical assessment. First, we can see a clear distinction between the tech-enhanced, smart living, located primarily in the city centre and the support services, such as those that maintain the data infrastructure, placed mostly in the periphery. Would the neighbourhoods that house the new data centres become smart, too – and what would that entail? Second, the environmental impact on the suburbs and the periphery is unclear. Data centres are mostly perceived as sites of sustainable energy (Velkova 2016). And yet, the actual benefits (or risks) of data centres for communities are rarely discussed, especially when placed in poor and disenfranchised neighbourhoods, where the residents have no bargaining power against infrastructure initiatives and plans that are framed as 'job opportunities' and 'affordable energy'.

These, of course, would be different for each city in question. Our discussions of Helsinki and Manchester do not aim to offer a single model of geographic analysis; rather, we call for a more critical and thorough consideration of how the geography of smart city visions relates to local geographies – both physical and social – and where does the environment figure when these geographies are imagined or challenged. We must ask, what kind of smart city presence is established where: service provision (in the form of data centres) in some deprived areas, nothing in others, smart bikes in the city centre, lack of public transport infrastructure in the periphery, and so on. Understanding the process of smartification as a geography of injustice also helps better understand why some of the initiatives might fail, why they were met with resentment and resistance from the residents, or why they were deprioritised by planners. And finally, beyond the brokenness of local geographies, smart cities also need to be explored globally, as part of digital economies' supply chains (Brevini 2021, 2023a) – the latter are almost aways invisible, for smart city imaginaries often deploy the logic of *out of sight, out of mind.* Just as smart cities' visions are often disconnected from the actual material places, landscapes, and communities, they also appear disconnected from the global supply chain; from the fact that they are contingent on resource extraction, labour, and pollution that take place somewhere else. How can we account for smart cities' environmental impact, when the brokenness they inflict takes place somewhere else, invisible to the eye and beyond the scope of one's imagination?

### Broken temporalities

Throughout the book we have shown that there are multiple temporalities at work beyond the dominant timeline of smart city – a linear progressive temporality defined by economic growth and technological innovation drive (see also Kitchin 2019).[2] These multiple temporalities include how smart city's imaginaries inherit city models of the industrial revolution. In using the Woolworth building and the Corbusian cities as examples, in Chapter One we suggested that smart cities follow modernist urbanism's understandings of nature as a resource that needs human management, cultivation, and control. Contrary to the claim of its novelty – its radical break from previous urban planning and urban design approaches – smart cities' promise to integrate social and economic concerns with environmental measures such as optimising energy usage, mitigating pollution and increasing green areas in urban spaces are not new. Rather, they are shaped by, and conditioned upon histories of human-technology-environment relations. This includes but is not limited to the infrastructure, materials, and labour that produce heat and electricity, the ecosystems that the smart districts are embedded in, the soil upon which the smart buildings are built, as well as the relations of extraction that take place outside the boundaries of smart cities in both spatial and temporal terms. The relations are also shaped by the temporality of the digital itself, especially its promise of 'real-time' information that is seen to hold the unique affordance for solving an increasingly long list of environmental concerns, such as carbon emissions, waste, energy efficiency, energy transition, and climate change.

The real-timeness of digitisation often goes hand in hand with acceleration (Virilio 2000; Wajcman 2014). In case of smart cities, digital technologies promise that one can move faster, and in a more efficient way (Kitchin 2019) – to the point of actually acquiring extra time, as we could see for example in Chapter Two, in the Kalasatama's promise that its residents would gain 'one more hour a day'. However, behind the promise of more and faster time lies the fragmented, broken temporality of smart cities' project economy. As we could see in both Helsinki and Manchester, smart city promises consist of various timelines and rhythms of funding schemes and projects, granted by academic or governmental bodies for a specific and limited period, that might begin and end at different times. When the funding ends, so does the project – and despite being connected by the overall discourse of smart cities, there is often little actual connection between the projects, across cities and even within each city. What we saw in both Helsinki and Manchester smart city initiatives can be called a *broken continuity* – a form of brokenness partly though not exclusively related to broken temporalities. As we looked at the number of various disjoint projects, we noticed that even when they are mentioned together (for example, in presentations or reports), they are discrete and not connected. In Manchester, in particular, there was no sense of a shared legacy, despite a number of projects taking place over more than a decade, often with the same institutions involved, such as universities or the City Council. There was also no continuity of information or insights, and that despite the emphasis on future benefits and explicit promises of creating a legacy that would last.

Another reason for this broken continuity is that despite taking place in one city, and despite a singular narrative or vision of a smart city that may be promoted by a national/ local policy or by corporate branding strategies, the reality of a smart city consists of small, localised, and often narrowly focused initiatives. Helsinki and Manchester are not unique in that respect: as Robert Cowley and Federico Capriotti (2019) note in their discussion of smart cities as 'anti-planning', the '"smart" concept rhetorically co-articulates an array of otherwise heterogeneous and often spatially dispersed projects, programmes of initiatives, promotional documents, and strands of official policy' (2019: 434). Smart cities, they continue, exist between idealised narratives and actual local conditions of built environment and local governance, into which they are 'awkwardly integrated' (Shelton et al 2015: 14, cited by Cowley and Capriotti 2019: 434).

Where does this broken continuity leave the environment? The 'anti-planning' nature of smart cities might help explain their narrow vision when it comes to environmental footprints of digitisation, even when such footprints are known by many of the people involved. But what often escapes the critique of smart city planning is the fact that environmental harms are built into the very concept of smart city due to its broken and disjoint temporalities. First of all, the complicated relation between the short-term project economy and the promise of sustainability that is often considered long-term is one that requires closer examination. On the one hand, the short duration of smart city projects means that any integrated, systematic, and long-term plan and implementation is impossible, not to mention tracking whether they have delivered their promises or measuring their environmental impacts. On the other hand, the long-termness of sustainability goals is often used to justify the limitations of the short-term projects. In Kalasatama the short project funding duration is seen as necessary for gaining more data and for testing out different technologies for achieving the best possible solution with long-term implications.

However, the temporal disparity between the fast production of data, such as the real-time air quality index, and its ellipses, disposability, and ruination, detailed in Chapter Two, cast into doubt the promise of digital solutions for environmental problems. Furthermore, the smart city's framework of long-term sustainability follows an economic logic that elides crucial issues such as the whole life-cycle environmental impact of materials and the path-dependencies and lock-ins of energy transition. For example, the high-rise buildings in Kalasatama are considered energy efficient. Yet, this evaluation excludes considerations of the environmental impact of the whole process of production, transportation, and disposing of construction materials, as well as of the environmental damage caused by burning biomass that generates heat for these buildings.

Second, and perhaps most important, the broken and fragmented temporalities of anti-planning, coupled with the fast-moving temporality of innovation which lies at the heart of smart city projects, exist in stark contradiction to the slowness of their environmental impact. What we are referring to here is the deep time of resources such as minerals or water utilised in manufacturing smart devices and sustaining digital infrastructures: while their extraction is both fast and relentless, their own temporality is extremely slow. We are also referring to the slowness of waste and e-waste decomposition generated by every project, whether it is immediately visible and momentarily sensationalised – as, for example, were the broken bikes seen in Manchester – or, more often, overlooked, ignored, and detached from the smart cities, by virtue of geographic distance, or the immenseness of the global digital supply chain. These and many forms of slow impact are part of what Rob Nixon has famously coined the 'slow violence' of environmental degradation that occurs over time and often out of sight (Nixon, 2011). Compared to the fast and the immediate, he notes, slow violence is hard to document since it takes a long time and demands a 'different ethical attention span' (2011: 211), and is even harder to communicate, as it is hard to 'convert into dramatic form urgent issues that unfold too slowly to qualify as breaking news' (2011: 210). Indeed, we are yet to see a well-funded *longitudinal* project, investing in, and exploring the impact of smartness of cities *over a long period of time*. How, then, can we account for what breaks, especially when it breaks very slowly?

## What remains?

We began our discussion of broken promises by thinking about *what breaks*. Being regularly confronted with a positivistic critique of our work, asking 'can you prove that it is broken?', we challenged the very assumption that it is always possible to account for, never mind quantify and measure, the impact of brokenness since, as we have shown, the brokenness often happens in a different place and time – somewhere else, invisible to the eye, or too slowly, and thus hard to see. We conclude our discussion with the question about *what remains*, what is left after – an aspect of innovation and experimentation that is very rarely addressed in visions of digital and smart living and techno-urbanism. Material brokenness, waste, and remains are excluded from the field of imagination and discursive possibility when the focus is on what *will* be created, what *will* be new, what *will* be more efficient. Similarly, digital waste, decay, and ruination are abundant in smart city projects but are rarely considered in the present- and future-oriented trajectories of smart city visions. The fast-moving speed of experimentation and data-driven real-timeness, seen in Helsinki (and observed in many other smart city frameworks, see Kitchin 2014), or the 'future-proofing' of cities to become fully digital, to use the words of Manchester's latest digital strategy (see Chapter Three), afford no room to consider what might need to be demolished to build the new; or what would be left behind when the new breaks down, is discarded, or becomes deactivated.

Digital brokenness of smart cities can take many shapes and forms. For example, it is about the brokenness of data – disused or inaccessible data sets; excess of passive data; and data that is simply discarded without delivering its promise (Savolainen 2023).[3] But perhaps the most visible examples are project websites and social media profiles. All smart city projects have at least one project website; in many cases, they are accompanied by extensive social media activity on platforms such as Facebook, YouTube, Twitter, or Instagram. Digital presence of smart city projects – a mandatory tool of public dissemination and audience engagement – is often structured around real-timeness: updates about project development; interactive data visualisations; maps and videos; and links to smart city publicity events such as launches, conferences, celebrations, blogs, or podcasts. Over time, especially as projects' funding ends, or as certain activities cease, this digital presence begins to break down. In some cases, we see websites that are no longer updated – as the example of Kalasatama website, described in Chapter Two – but remain online as a sort of virtual ruin. In other cases, digital ruination, just as material ruination, also includes slow crumbling and decay, when the project façade remains accessible but various components begin to disappear. This was clearly seen in the case of Manchester smart city projects that took place in mid–late 2010s, which were discussed in Chapter Three. Several years after the projects have ended, their websites were also gone – however their social media accounts remained. Walking/scrolling back through Facebook and Twitter was akin to walking in the ruins of a mostly preserved building, letting one's imagination do the work of recreating (or making up) the missing details. Social media accounts of Manchester's two smart city initiatives allowed seeing a record of their past real-timeness – regular posts, links to frequent project podcasts and off-line events – but none of the links were working, since all had been hosted on the projects' websites. Apart from the frustration of not being able to find more information about the projects' activities, digital brokenness also poses a question of material impacts of digital ruins. Digital ruins do not exist in virtual domains alone: data storage and website hosting have real environmental implications, including a continuous carbon footprint (Greenwood 2021).

Material brokenness of smart cities and their waste and remains can similarly take many forms. These include the more visible and somewhat sensationalised broken bikes; the semi-visible ruins of older architectural structures that need to be taken down for new, more sustainable alternatives to be built in their place; and the largely invisible e-waste – an electronic graveyard, inevitably generated at the end of various project life cycles, as well as during routine digital consumption. It is a graveyard that usually takes place *elsewhere*, outside the smart city perimeters, and often on a different continent: as scholars of e-waste remind us, both recycling and disposal of electronics from the so called 'developed' world take place in the so called 'developing world' (see for example Gabrys 2011). Despite the emphasis on sustainability, most digital devices, whether those developed specifically as part of smart city initiatives, or those in general circulation (such as smartphones), on which the participation in smart urban living is usually contingent, are disposable by design (Chen 2016). Sensors and smart lights that have completed their role in an experiment and are no longer required; old equipment left behind after upgrades to more sustainable alternatives;[4] and, of course, the smart city residents' own personal devices. Every city service that turns digital, that requires a newer smartphone, or another smartcard, will lead to more digital consumption and, as a result, more e-waste, both individual and institutional.

Focusing our attention on ruins – both digital and material – allows us to bring to the fore the notion of the 'aftermath'. Centring 'the aftermath' rather than as a more generic and amorphous category of 'future', we argue, offers a more grounded and more precise discussion of smart cities and their environmental impact; and especially of what happens to them in the long durée, beyond the various short-term experiments that come and go. The *aftermath* allows us to move past being caught up in the specific promises – of extra time, of cleaner energy, of more sustainable development, of interconnected data – and from specific evaluations of whether and to what extent these have been delivered. Such evaluations have a danger of simplifying a complex problem to a question of measurement, metrics, and project management, all of which are by definition localised rather than embedded in national, regional, and international flows of capital and materials; and are practical rather than critical. They do not, and cannot, account for the global matrix of digital supply chains, the complexity of global climate injustice, or the invisible yet tangible matrix of racial and colonial legacies that shape seemingly neutral, technocratic visions of 'sustainable futures'.

Furthermore, focusing on the *aftermath* allows us to consider what else can happen to a promise, beyond being delivered or broken, and to imagine the life of broken promises over time, past the immediate aftermath of a project. As noted earlier in this chapter, to account for the slow violence of environmental degradation and the role of digital technologies in it, we need to turn our attention to what might be breaking very slowly, in a way that escapes any short-term evaluation. Such an epistemological slowing down will require a shift in how we think about futurity, and about the life trajectory of harms: instead of directing our gaze backwards, at what was broken/harmed, or at which ruins of the past are haunting the present, we should turn to the future. Inspired by the notion of haunted futurities as a form of responsibility (Ferreday and Kuntsman 2011), we should ask: what kind of futures can we imagine in the ruins of a promise? And what kind of responsibilities towards the future do we have, if we aim to undo the violence of cruel optimism embedded in ideas of smart sustainability and digital innovation?

## Caring for a future in the ruins of broken promises

One of the key aims of the book was to decentre the digital in the imagination, narrative, and practices of smart cities. To do so we developed the theoretical and methodological framework of broken promises to situate our analysis in the ruins of smart cities – between the digital and the environment. Ruins are time binding. They testify to the multiple spatial-temporal relations that take shape through fissures, frictions, and gaps. They embody the historical sedimentations, material, and symbolic transformations, practices of production, abandonment, and their aftermath. Our analysis of the ruins calls into question the assumed spatial and temporal boundaries of smart cities. It teases out their complex and contradictory configurations, as observed through the gap between what is promised and what is delivered, as well as what the expected, justified, and valued failures are. Our book invites the reader to look for the things and relations that break, projects that get discontinued, and environmental and social harms that are rendered *out of sight and out of mind* and which therefore cannot be immediately seen or made intelligible. In doing so, we also call for attention to the power structures that shape and sustain these harms' invisibility, for they are not invisible or unintelligible for all, least of all to those who live in their proximity. Or, as Thom Davies in his discussion of slow violence and toxic geographies poignantly asks, 'out of sight for whom?' (Davies 2022; see also Alexis-Martin and Davies 2017).

Shifting away from specific promises to broader questions of long-lasting-ness, harms, and brokenness allows to foreground questions of accountability, as we attempt to decentre the digital from the conversation about smart cities and environmental justice. We suggest that such a conversation should be forward rather than backward looking; in other words, instead of assessing the harms after they have taken place, and seeking to either blame or repair, we should *begin* the conversation with asking about potential harms. Instead of simply asking what is currently broken, we should shift the conversation about smart cities' futures into what *might*, and *would* break, when digitisation is driven by short-term, disjoint experiments, when it is based on the logic of experimentation and project economy, and when environmental questions are separated from digital agendas, or deprioritised within them. Our focus on ruins thus offers a refreshing entry point: when we are observing digital ruination in real time, wondering how long the current digital and material objects will last, we are already anticipating the brokenness; and already historicising the ruination, leaving behind the myopia or short-sighted project cycles and real-timeness of the now.

Historicising the future helps us bring into context issues that might not be immediately considered as linked, for example I(o)Ts and the $CO_2$ emission and biodiversity loss in the next decades and centuries as a result of burning biomass; the production of 'real-time' environmental data and the endless processes of digital ruination; the agile project economy and the slow violence inflicted on the environment and the perpetual structural inequalities bodily experienced by the marginalised communities. Thinking about what might happen *in the ruins of broken promises* offers a theoretical language that enables us to simultaneously historicise *and* reimagine the future of smart cities, and the future of human-technological-environmental relations more broadly, away from the dominant narratives and imaginaries that rest on the cruel optimism of techno-solutionism and digital inevitability.

What, then, can one look for in the ruins of broken promises? One possibility would be to reorient the approach to smart technologies from solutionist to what some scholars have coined 'care-ful'. For example, in the recently published special issue on 'care-ful data studies', the editors, Irina Zakharova and Juliane Jarke, ask: 'what do we see, when we look at datafied societies through the lens of care' (Zakharova and Jarke 2024: n.p.). To answer this question they offer a number of principles to guide future feminist scholarship on data, among them a move from critique to care and responsibility which includes attention to materialities, distributions of resources, continuity, and repair; a move from studying harms to focusing on vulnerabilities and care; and, most importantly, a shift from data-driven technologies to socio-digital care arrangements.

The simultaneous attention to damage and repair, accountability for harms, and capacity to imagine alternatives is incredibly inspiring for thinking differently about the relations between digital technologies and environmental care. And yet, such care-ful approach to data – and, by extension, to digital technologies more broadly – still revolves around the digital and the hope in its potential (as long as its use is care-driven). As we dwell on the unfolding effects of slow violence inflicted by digitisation, we wonder, instead, about the possibilities of care embedded in the call for digital disengagement (Kuntsman and Miyake 2022) – an approach that challenges the normalised nature of digitisation in all spheres of today's life. Digital disengagement points to the potential of disconnection, logging out, opting out, and refusing digital solutions, but not as an individual act of temporarily leaving one's phone behind or reducing screen time. Rather, digital disengagement is a collective form of resistance against social, racial, economic, and indeed environmental forms of injustice brought on by enforced, and constantly expanding digitisation. As the authors of *Paradoxes of Digital Disengagement: In search of the Opt Out button* put it:

> we are calling for the decoupling, denaturalisation, and destabilisation of the digital as the starting and ending point for all. We do not need to reject scientific progress, or ignore the usefulness of digital media and technologies where it creates new access, opportunities, and solidarity tools for those fighting oppression. But what we do need is a fundamental change of perspective in how we think about digital technologies, as a synonym of desired futurity. Each time we imagine or plan a future, instead of considering digital solutions as the default option, we should undo the metonymic connection between 'futures' and 'digital'. Rather than asking, how should a particular new technology, device or platform be designed, governed and used, we should ask instead: what are its consequences? Where are the possible ways out of this digital plan for those whom it may not fit? And most importantly, what are the alternatives? (Kuntsman and Miyake 2022: 151)

In the context of environmental harms, digital disengagement might look like a turn towards non-digital pathways for environmental care and human-nature relations in urban life – for example, via community-based food cooperatives and local food production that is based on principles of degrowth, slowing down, low tech, and agroecology.[5] It might look like a call to 'reduce, reuse, refuse' the digital itself[6] – a radical rethinking of digital consumption and use that draws on the principle of 3R (reduce/reuse/recycle), sometimes also formulated as 5R to include repurpose and refuse. Going beyond individual acts such as buying second-hand phones or carefully recycling electronics, the principle of multiple Rs in the context of digitisation challenges the very foundation of digital political economy, by substituting the logic of capitalist expansion with environmentally driven degrowth and decolonial and anti-racist environmental justice. At the heart of these and similar alternatives lies what Michelle Murphy (2015) calls the practice of unsettling care, aiming to both dismantle and rebuild human-technology-environment relations, beyond simplistic or innocent notions of care-with-the-digital, and perhaps even beyond the 'care-ful' scholarship of digital technology. Murphy's call for unsettling care troubles the technoscientific timescape of futurity, making visible and reassembling practices and relations that have been neglected in the dominant narratives.

Although not explicitly focused on digitisation or smart cities, we find Murphy's formulation particularly inspiring, as we try to decouple futurity and environmental care from the digital. As we look for justice and hope *in the ruins of broken promises* and ask how they can be unsettled? To unsettle the promises that are *already* broken means acknowledging that human-technology-environment relations are multiple and result in a history of configurations, arrangements, and alterations whose ramification unfolds in ways that are both felt and unintelligible, both fast and slow, both emergent and continuous, both local and planetary. It also means moving away from the accelerated speed of digital capitalism (Virilio 2000; Wajcman 2014) and from the temporality of immediacy – be that of the immediacy of the digital, or even the immediacy of trying to fix that which does not work. In the place of immediacy, unsettling with care would prioritise slow and long-term repair, sedimented relations, and what Kate Cairns coined 'pedagogies of excavation' (Cairns 2021) – a way to slowly and gradually understand and explore layers and histories that are underpinning environmental injustices, and resist them in ways that care for, and prioritise knowledge and experiences of communities who are directly affected by structures of social and environmental violence.[7]

Reassembling, rearranging, excavating, and repairing slowly! This is our redefinition of innovation and futurity. The reassembling and rearranging of the multiple temporalities and spatialities that constitute the broken promises of smart cities allows for imagining and constructing human-technology-environment relations beyond the techno-optimistic marketing dream of being – or becoming – digital and smart. And excavation and slow repair create space where ruination and digital environmental harms can be resisted and undone – over time and for many generations to come.

# Notes

## Introduction: How Do We Think About Smart Cities?

[1] By district, we refer here to an area within a city, rather than within a country. For example, Kalasatama is described as a smart district of the city of Helsinki.

[2] Knox (2020) notes that in the present moment, calculating carbon emissions is the main way to make climate change intelligible. Donaghy et al (2023) push this point further by arguing that current discourses around climate change are 'carbon centric', obscuring other forms of environmental injustice.

[3] Throughout the book, we use the notion of environmental care, drawing on feminist perspectives of care (Murphy 2015; de la Bellacasa 2017).

[4] Throughout the book, we use the term 'project' in two ways. The first one refers to smart city as an abstract idea, a concept that is linked to matters of governance, economy, and planning. The second is about concrete local endeavours that have a clearly defined focus (that is, to develop a 'smart district', or to create a technological innovation for one city) and are usually funded by research councils, governments, or industry for a specific time period.

[5] For a specific discussion of digital fetishism, see Miconi (2023). See also Fuchs (2019) for a broader discussion of Marxist approach to capitalism in times of big data.

[6] Benedetta Brevini's (2021) work on AI and its environmental harms is one of the few pioneering exceptions.

## one  Smart Cities, Digitisation, and the Environment

[1] While our analysis is not explicitly situated within the science and technology studies (STS) literature, our argument here echoes much of the STS work on the role of science, and of digital technologies in particular, in producing knowledge about the environment and the climate (see for example Calvillo 2023; Knox 2020; Noortje 2015).

## three  Manchester

[1] It is also not a full-scale ethnography of the many actors involved in the process of making Manchester a smart city. Such an ethnography, akin

to earlier work by Hannah Knox (2020) on tackling climate change in Manchester, would undoubtedly be beneficial to fully understand how 'smart', 'digital', and 'environment' are understood, constituted, and lived with.

2    The interviewees were from the Manchester City Council team; the City Council's Digital Strategy Team; Open Data Manchester, a non-profit organisation 'supporting people, communities and organisations to understand and use data that is about them and their communities' (https://www.opendatamanchester.org.uk/); and two environment- and climate-focused organisations: a North-West-based organisation supporting access to nature, and a Manchester-based organisation, supporting climate change action, which requested to be anonymous.

3    The short trailer of the film can still be seen on City Verve's YouTube channel (YouTube 2018), as well as on many websites and news articles, where the clip was embedded. The full film, which was hosted on City Verve's website, is no longer available.

4    Professor John Davies, Lead Researcher at BT.

5    Many of the reports used to write this chapter were found on-line; a few were provided by a University of Manchester researcher and a City Council employee, both of whom were involved in the smart city projects.

6    WayBack Machine archives the pages by timestamps of their capture: on certain dates, certain parts of the website might be available – or not.

7    The projects' disappearing *digital* legacy could be read in many ways. One is the perspective of software studies, focusing on legacy systems as a matter of computer order processing, where past persistence and 'felt absence' is understood as 'material-semiotic problems of appointment and designation, sending and receiving' (Stevenson and Helmond 2020: 2). Another is the perspective of digital memory, where our attention is turned to platform affordances and governance and their role in shaping what is kept and what is lost; what can be forgotten and what will be remembered. As the focus of this chapter is more on the legacies of smart city projects – and in particular, on their environmental impact – my theorising of digital legacy is deliberately left short, as expanding on its theoretical complexity lies beyond the scope of this book.

8    Of course, what is understood as being 'on target' requires a more detailed exploration, which is beyond the scope of this chapter. For an excellent anthropological exploration of the role of numbers in thinking about, and acting on climate change in Manchester, see Knox's *Thinking Like a Climate*, especially as she seeks to 'unravel why it has come to make sense to govern climate change in terms of percentage reductions in carbon emissions, and how this approach relates to the way in which climate change has been established as a problem by climate science' (2020: 44).

[9]   Questions were adapted to the interviews with those working with smart city projects versus those working in climate- or environment-focused organisations. For the latter, the questions sent ahead of time were as follows:

1. How does your organisation see Manchester from the point of view of climate change and environmental concerns? What are the main issues affecting people in Manchester?

2. Manchester is an aspiring smart city. How do you think the development of Manchester as smart city will impact the environment and climate change? Does your organisation have a position/policy/ activity planning specifically related to smart cities (in Manchester and perhaps elsewhere in the region)?

3. What is your awareness of carbon footprints of the digital/smart technologies used in Manchester? Please elaborate.

4. What is your awareness of e-waste footprint of digital technologies used in Manchester? Please elaborate.

5. What is your awareness of other environmental issues related to digital technologies involved in making Manchester a smart city?

6. What is your awareness of which countries and communities are most harmed by the digital technologies manufacture, use, and disposal?

7. What kind of environmental sustainability measures should Manchester have in place, in your view?

8. How can your organisation help minimise environmental and climate harms of digitisation?

9. What else, in your view, can be done to minimise environmental degradation and climate crisis in Manchester and globally?

[10]  Heat emissions of data centres is a complex issue that is discussed extensively, both as a form of sustainable and effective energy use and as a form of environmental footprint (see Velkova 2016).

## Conclusion: In the Ruins of Broken Promises

[1]   As the report explains, 'Experimentation has provided lessons and knowledge about new services for the future that are not yet on the market. Among other things, the pilot projects have tested how new services are accepted, how they function and how they should be developed further. Information that identifies future needs will benefit both the city's designers and companies developing services' (2021: 14).

[2]   In his article on the timescape of smart cities, Rob Kitchin noted that they often create multiple temporal rhythms, relations, and modalities (Kitchin 2019). Our own analysis of the two smart cities resonates with this observation, as we elaborate on later. However, the framework of

broken promises offers an additional insight. Rather than seeing multiple temporalities as neutrally co-existing, we argue that they should be understood as discrepancies that generate environmental harms and socio-economic injustices.

3   The reasons for data brokenness are complex and might do with the end of an experiment where the data collected is no longer needed; or custodianship of data which is tied to research funding and not shared publicly; or corporate data ownership which prevents public access.

4   One of Adi's interviewees who works with Manchester City Council discussed the need to reduce the footprint of the Council's own estate and employees, including the computer equipment used – a very valuable point, unusually attentive to the footprint of digital working. Yet, what remains invisible and under-explored, even in discussions like this, is the fate of the old equipment once it is replaced. Indeed, several of the interviewees noted that the facilities for safely disposing and recycling e-waste in Manchester are extremely limited.

5   A Manchester-based Mushroom co-op is an excellent example of such practice (Myco n.d.)

6   Despite the popularity of this principle in the general context of ethical consumption, it has rarely been applied to digital technologies. One of the first attempts to explore it took place in an online summer school dedicated to digital technologies and the environment (Digital Politics 2023).

7   For example, pedagogies of excavation 'work to unearth the violent systems that create these conditions [of environmental harms]: to connect polluted air and food apartheid to the relations of racial capitalism' (Cairns 2021: 538).

# Bibliography

6Aika (2015) *The Six City Strategy – Open and Smart Services*, https://6aika.fi/wp-content/uploads/2015/11/6Aika-strategi a_p%C3%A4ivitys_2015_EN.pdf, accessed 26 September 2023.

Acea Waidy Wow (n.d.) 'Smart city: what is it and what are its characteristics', Acea Waidy Wow, https://waidy.it/en/green-solutions/smart-city.html

Ahmed, S. (2004) *The Cultural Politics of Emotion*. Edinburgh: Edinburgh University Press.

Ahmed, S. (2012). *On Being Included: Racism and Diversity in Institutional Life*. Durham: Duke University Press.

Akese, G. and Little, P. (2018) 'Electronic waste and the environmental justice challenge in Agbogbloshie', *Environmental Justice*, 11(2): 77–83.

Alexander, D. (2021) 'Andy Burnham urges residents to not chuck hire bikes in canals', *Road CC*, 13 November 2021, https://road.cc/content/news/andy-burnham-asks-residents-not-chuck-bikes-canal-287769, accessed 1 July 2023.

Alexis-Martin, B. and Davies, T. (2017) 'Towards nuclear geography: Zones, bodies, and communities', *Geography Compass*, 11(9): 1–13.

Álvaro-Alonso, I. M. (2021) 'Data and digitalisation, two key tools in the fight against climate change', *Telefónica*, 9 June 2022, https://www.telefonica.com/en/communication-room/blog/data-and-digitalisation-two-key-tools-in-the-fight-against-clim ate-change/, accessed 1 August 2023.

Ameel, L. (2021) *The Narrative Turn in Urban Planning: Plotting the Helsinki Waterfront*. New York: Routledge.

Andreucci, D., García-Lamarca, M., Wedekind, J. and Swyngedouw, E. (2017) '"Value grabbing": A political ecology of rent', *Capitalism Nature Socialism*, 28(3): 28–47.

Aouragh, M., Gürses, S., Pritchard, H., and Snelting, F. (2020) 'The extractive infrastructures of contact tracing apps', *Journal of Environmental Media*, 1(1): 9.1–9.9.

Arnold, D. (2005) 'Europe, technology and colonialism in the 20th century', *History and Technology*, 21(1): 85–106.

Association for Project Management (n.d.) 'Triangulum – Manchester moves towards smart zero carbon', https://www.apm.org.uk/resources/find-a-resource/case-studies/case-study-manchester-moves-towards-smart-zero-carbon-with-triangulum/

Baran, Z. and Stoltenberg, D. (2023) 'Tracing the emergent field of digital environmental and climate activism research: A mixed-methods systematic literature review', *Environmental Communication*, 17(5): 453–468.

Barlow, N. (2018) 'Manchester-A City of Firsts celebrates the city's world changing innovations', *ABOUT Manchester*, 7 June 2018, https://aboutmanchester.co.uk/manchester-a-city-of-firsts-celebrates-the-citys-world-changing-innovations/, accessed 5 March 2024.

Barr-Engel, S., Cook, S., Cullen, C., Donti, P., Hopps-Weber, S., Jordy, A., Julius, A., Krouse. C., Loftus, F., Maharaj, T., Minns, C., Mocorro, A., Olajde, O., Paul, S., Sanders, J., Selbe, S., Carlo, E. S., Spratling, D., and Vargas, C. (2021) *Environmental Justice in Technology Blueprint, Earth Hack and Partners*, https://drive.google.com/file/d/1hpqCcT2Pp74L4Cx3NeErBCYf7dfflaPH/view, accessed June 2024.

Barsanti, M., Garbolino, L., Mansoor, M., Realmonte, G., Zeinoun, R., Causone, F. and Fabi, V. (2020) 'Innovative user experience design and customer engagement approaches for residential demand response programs', in J. Littlewood, R. J. Howlett, A. Capozzoli, and L. C. Jain (eds) *Sustainability in Energy and Buildings: Proceedings of SEB 2019*. Singapore: Springer, 613–627.

Basiago, A. D. (1996) 'The search for the sustainable city in 20th century urban planning', *The Environmentalist*, 16: 135–155.

Benjamin, R. (2019) *Race After Technology: Abolitionist Tools for the New Jim Code*. Cambridge: Polity.

Berlant, L. (2010) 'Cruel optimism', in M. Gregg and G. J. Seigworth (eds) *The Affect Theory Reader*. Durham: Duke University Press.

Berlant, L. (2011) *Cruel Optimism*. Durham: Duke University Press.

Blair, J. J., Balcázar, R. M., Barandiarán, J., and Maxwell, A. (2023) 'The "afterlives" of green extractives: Lithium mining and exhausted ecologies in the Atacama Desert', in F. Filipe Calvão, M. Archer, and A. Banya (eds) *The Afterlives of Extraction: Alternatives and Sustainable Futures*. Leiden: Brill.

Bonneuil, C. and Fressoz, J.-B. (2016) *The Shock of the Anthropocene: The Earth, History and Us*, translated by David Fernbach. New York: Verso.

Borsboom-van Beurden, J., Kallaos, J., Gindroz, B., Costa, S., and Riegler, J. (n.d.) *Smart City Guidance Package: A Roadmap for integrated planning and implementation of smart city projects*, Marketplace of the European Innovation Partnership on Smart Cities and Communities: A consultation paper to stimulate action. European Innovation Partnership no Smart Cities and Communities.

Brevini, B. (2020) 'Black boxes, not green: Mythologizing artificial intelligence and omitting the environment', *Big Data & Society*, 7(2), 205395172093514.

Brevini, B. (2021) *Is AI Good for the Planet*. Cambridge: Polity.

Brevini, B. (2023a) 'Myths, techno solutionism and artificial intelligence: Reclaiming AI materiality and its massive environmental costs', in S. Lindgren (ed) *Handbook of Critical Studies of Artificial Intelligence*, Cheltenham, UK: Edward Elgar Publishing.

Brevini, B. (2023b) 'Artificial intelligence, artificial solutions placing the climate emergency at the center of AI developments', in H. S. Sætra (ed) *Technological Change, Society, and Sustainability*. New York: Routledge.

Brevini, B. and Murdock, G. (eds) (2017) *Carbon Capitalism and Communication: Confronting Climate Crisis*. Palgrave Studies in Media and Environmental Communication. Cham: Springer International Publishing.

British Telecom (2017) 'Connecting Manchester: How BT's Internet of Things solutions became central to the CityVerve smart city project', BT Report, https://www.iot.bt.com/assets/documents/bt-city-verve-smart-city-report.pdf, accessed 20 January 2023.

Caird, S. (2018) 'City approaches to smart city evaluation and reporting: Case studies in the United Kingdom', *Urban Research & Practice*, 11(2): 159–179.

Cairns, K. (2021) 'Feeling environmental justice: Pedagogies of slow violence', *Curriculum Inquiry*, 51(5): 522–541.

Calder, B. (2022) *Architecture: From Prehistory to Climate Emergency*. London: Pelican.

Calvillo, N. (2023) *Aeropolis: Queering Air in Toxicpolluted Worlds*. New York: Columbia University Press.

Caprotti, F. (2014) 'Eco- urbanism and the eco-city, or denying the right to the city?' *Antipode*, 46(5): 1285–1303.

CB Insights (2020) 'What are smart cities', *Research Briefs,* December 15, 2020, https://www.cbinsights.com/research/what-are-smart-cities/

Chen, S. (2016) 'The materialist circuits and the quest for environmental justice in ICT's global expansion', *TripleC: Communication, Capitalism & Critique. Open Access Journal for a Global Sustainable Information Society*, 14(1).

City of Helsinki (2020) *Environmental Report 2019*, https://julkai sut.hel.fi/sites/default/files/2020-09/helsinki-environmental-rep ort-2019-web.pdf, accessed 26 September 2023.

City of Helsinki (n.d.) *Kalasatama: A multicoloured side of the city*, City of Helsinki, City planning department, https://www.hel.fi/sta tic/hel2/ksv/julkaisut/esitteet/kalasatama_esite_en.pdf, accessed 3 August 2023.

City Verve (n.d.) 'Seeing the benefits of bicycle sensors', *Internet Archive WayBack Machine*, https://web.archive.org/web/2017090 5225607/http:/www.cityverve.org.uk/seeing-benefits-bicycle-sensors/, accessed 3 May 2024.

Collier, C. (2019) 'The Manchester model: How a city's urban past influences its smart city future', *Smart Cities Connect: Media and Research*, https://smartcitiesconnect.org/the-manchester-model-how-a-citys-urban-past-influences-its-smart-city-future/, accessed 5 June 2023.

Corbusier, L. (1929) *The City of To-Morrow and its Planning*, translated by Frederick Etchells. New York: Dover Publications.

Cowley, R. and Capriotti, F. (2019) 'Smart city as anti-planning in the UK', *Society and Space*, 37(3): 428–448.

Cowley, R., Joss, S., and Dayot, Y. (2018) 'The smart city and its publics: Insights from across six UK cities', *Urban Research & Practice*, 11(1): 53–77.

Coyle, S. (2018) 'Dumped in canals and hung on traffic lights – what people have done to Mobikes in Manchester', *Manchester Evening News*, 5 September 2018, https://www.manchestereveningnews.co.uk/news/greater-manchester-news/gallery/dumped-canals-hung-traffic-lights-15114162, accessed 1 June 2023.

Cugurullo, F. (2021) *Frankenstein Urbanism: Eco, Smart and Autonomous Cities, Artificial Intelligence and the End of the City*. Abingdon, UK: Routledge.

Dastbaz, M., Naudé, W., and Manoochehri, J. (eds) (2019) *Smart Futures, Challenges of Urbanisation, and Social Sustainability*. Cham: Springer International Publishing.

Davies, T. (2022) 'Slow violence and toxic geographies: "Out of sight" to whom?' *Environment and Planning C: Politics and Space*, 40(2): 409–427.

de la Bellacasa, M. P. (2017) *Matters of Care: Speculative Ethics in More Than Human Worlds*. Minneapolis: University of Minnesota Press.

Demarque, C., Charalambides, L., Hilton, D. J., and Waroquier, L. (2015) 'Nudging sustainable consumption: The use of descriptive norms to promote a minority behavior in a realistic online shopping environment', *Journal of Environmental Psychology*, 43: 166–174.

Department for Digital, Culture, Media and Sport (2015) 'Manchester wins £10m prize to become world leader in "smart city" technology', Press release, 3 December 2015, https://www.gov.uk/government/news/manchester-wins-10m-prize-to-become-world-leader-in-smart-city-technology

Di Salvo, A. L., Agostinhob, F., Almeida, C. M. V. B., and Giannetytti, B. F. (2017) 'Can cloud computing be labeled as "green"? Insights under an environmental accounting perspective', *Renewable and Sustainable Energy Reviews*, 69: 514–526.

Digital Politics (2023) 'Can we "reduce, reuse, refuse" the digital? Digital Politics Summer School, 5–6 June, 2023', *Digital Politics at Man Met*, https://digitalpoliticsmanmet.bloggi.co/can-we-reduce-reuse-refuse-the-digital-digital-politics-summer-school-5-6-june-2023, accessed 28 November 2023.

Digital Property and Cities (n.d.) *CityVerve: Evaluation and Recommendations*. ARUP. https://www.arup.com

Discard Studies (n.d.) https://discardstudies.com/

Donaghy, T., Healy, N., Jiang, C. Y., and Battle, C. P. (2023) 'Fossil fuel racism in the United States: How phasing out coal, oil, and gas can protect communities', *Energy Research & Social Science*, 100(103104): 1–16.

Earth Hacks (2023) *Environmental Justice in Tech Principles*, https://www.environmentaljustice.tech/ejit-principles, accessed 2 May 2024.

European Commission (n.d.) *European Regional Development Fund 2014–2020*, 22 June 2023, https://ec.europa.eu/regional_pol icy/funding/erdf/2014-2020_en

Fairbrother, M. (2016) 'Externalities: Why environmental sociology should bring them in', *Environmental Sociology*, 2(4): 375–384.

Ferreday, D. and Kuntsman, A. (2011) 'Introduction: Haunted futurities'. *Borderlands e-Journal*, 10(2): 1.

Forum Virium Helsinki (2021) 'Kalasatama has become known for agile pilots – the final report on the Smart Kalasatama project summarises the results of this long-term work', 16 November 2021, https://forumvirium.fi/en/publication/smartkalasatama-final-report/, accessed 8 November 2023.

Forum Virium Helsinki (n.d.) *Forum Virium Helsinki*, https://foru mvirium.fi/en/, accessed 10 July 2023.

Fraser, E. and Willmott, C. (2020) 'Ruins of the smart city: A visual intervention', *Visual Communication*, 19(3): 353–368.

Fraunhofer Institute IAO (2020) 'Triangulum: Demonstrate, disseminate, replicate', D2.6 Final Impact Report M60, January 2020.

Fuchs, C. (2019) 'Karl Marx in the Age of Big Data Capitalism', in D. Chandler and C. Fuchs (eds) *Digital Objects, Digital Subjects: Interdisciplinary Perspectives on Capitalism, Labour and Politics in the Age of Big Data*. London: Westminster University Press, 53–71.

Gabrys, J. (2011) *Digital Rubbish: A Natural History of Electronics*, Ann Arbor: University of Michigan Press.

Gabrys, J. (2022) *Citizens of worlds: Open-Air Toolkits for Environmental Struggle*. Minneapolis: University of Minnesota Press.

Gabrys, J. (2014) 'Programming environments: Environmentality and citizen sensing in the smart city', *Environment and Planning D: Society and Space*, 32(1): 30–48.

Galderisi, A. (2018) *Smart, Resilient and Transition Cities: Emerging Approaches and Tools for a Climate-Sensitive Urban Development*. 1st edition. Cambridge, MA: Elsevier.

Gale, F., Ascui, F., and Lovell, H. (2017) 'Sensing reality? New monitoring technologies for global sustainability standards', *Global Environmental Politics*, 17(2): 65–83.

Gassmann, O., Böhm, J., and Palmié, M. (2019) *Smart Cities: Introducing Digital Innovation to Cities*. Bingley, UK: Emerald Publishing Limited.

Giraud, E. H. (2019) *What Comes After Entanglement: Activism, Anthropocentrism and an Ethics of Exclusion*. Durham: Duke University Press.

Good, J. E. (2016) 'Creating iPhone dreams: Annihilating e-waste nightmares', *Canadian Journal of Communication*, 41(4): 589–610.

Gordon, A. (1997) *Ghostly Matters: Haunting and the Sociological Imagination*. Minneapolis: University of Minnesota Press.

Green Alliance (2020) 'Smart and green: Joining up digital and environmental priorities', Green Alliance, 20 October 2020, https://green-alliance.org.uk/publication/smart-and-green/ https://green-alliance.org.uk/wp-content/uploads/2021/11/ Smart_and_green.pdf, accessed 1 May 2024.

Green, B. (2019) *The Smart Enough City: Putting Technology in Its Place to Reclaim Our Urban Future*. Cambridge, MA: MIT University Press.

Greenwood, T. (2021) *Sustainable Web Design*. A Book Apart.

Gunder, M. and Hillier, J. (2009) *Planning in Ten Words or Less: A Lacanian Entanglement with Spatial Planning*. Farnham: Ashgate.

Gunster, S. (2022) 'Connective action, digital engagement and network-building: A Year in the life of Canadian climate Facebook', *Environmental Communication*, 16(5): 645–663.

Haarstad, H. (2017) 'Constructing the sustainable city: Examining the role of sustainability in the "smart city" discourse', *Journal of Environmental Policy & Planning*, 19(4): 423–437.

Hämäläinen, M. (2021) 'Urban development with dynamic digital twins in Helsinki City', *IET Smart Cities*.

Haraway, D. (1998) 'Situated knowledges: The science question in feminism and the privilege of partial perspective', *Feminist Studies*, 14(3): 575–599.

Haritaworn, J. (2020) '#NoGoingBack: Queer leaps at the intersection of protest and COVID-19', *Journal of Environmental Media*, 1(1): 12.1–12.7.

Härkönen, K., Lea, H., and Pyrhönen, O. (2022) 'Advancing the smart city objectives of electric demand management and new services to residents by home automation – Learnings from a case', *Energy Efficiency*, 15(25): 1–13.

Hasenörl, U. (2018) 'Rural electrification in the British Empire', *History of Retailing and Consumption*, 4(1): 10–27.

Helen Ltd (n.d.) 'Helen Ltd: About us', https://www.helen.fi/en/about-us, accessed 27 September 2023.

Helsinki Innovation Districts (n.d.a) Homepage, https://fiksukaupunki.fi/en/frontpage/, accessed 10 November 2023.

Helsinki Innovation Districts (n.d.b) 'Green Kalasatama', https://fiksukaupunki.fi/en/projects/green-kalasatama/, accessed 8 November 2023.

Hemment, D., Weise, S., and Conteh, F. (2018) 'Evaluation of human-centred design in City Verve: learnings for large scale demonstrators', FutureEverything Report, https://futureeverything.org/wp-content/uploads/2018/11/FE-HCD-in-CityVerve-REPORT-FINAL.pdf, accessed 5 March 2024.

Herrera, K. A., Pelegrin, J., Gayo, E., and Santoro, C. M. (2019) 'Circulation of objects and raw material in the Atacama Desert, Northern Chile by the end of the Pleistocene', *PaleoAmerica*, 5(4): 335–348.

Higginbotham, S. (2018) 'The internet of trash: IoT has a looming e-waste problem. A lack of forethought will leave us with a mountain of obsolete devices and no way to dispose of them', *IEEE Spectrum for the Technology Insider*, 17 May 2018, https://spectrum.ieee.org/the-internet-of-trash-iot-has-a-looming-ewaste-problem, accessed 1 July 2023.

Hird, M. (2022) *A Public Sociology of Waste*. Bristol: Bristol University Press.

Hollands, R. G. (2008) 'Will the Real Smart City Please Stand Up? Creative, Progressive or Just Entrepreneurial?' *City: Analysis of Urban Trends, Culture, Theory, Policy, Action*, 12: 303–320.

Howard, E. (1898) *To-morrow: A Peaceful Path to Real Reform*. London: Swan Sonnenschein.

Howard, P. (2015) 'Digital citizenship in the afterschool space: Implications for education for sustainable development', *Journal of Teacher Education for Sustainability*, 17(1): 23–34.

Hughes, M. (2020) 'Digital transformation: The key to tackling climate change', *Forbes*, 23 December 2020, https://www.forbes.com/sites/mikehughes1/2020/12/23/digital-transformation-the-key-to-tackling-climate-change/, accessed 1 September 2023.

Hyötyläinen, M. (2023) 'Land rent and the struggle for the urban commons in Helsinki's Suvilahti DIY skatepark', in H. HyHyötylötyläinen and R. Beauregard (eds) *The Political Economy of Land: Rent, Financialisation and Resistance*. New York: Routledge, 211–227.

IMD (2020) 'Singapore, Helsinki and Zurich triumph in global smart city index', September 2020, https://www.imd.org/news/competitiveness/updates-singapore-helsinki-zurich-triumph-global-smart-city-index/, accessed 12 November 2023.

Innovative Comms (2017) 'Cityverve', https://www.innovatecomms.co.uk/cityverve, accessed 15 March 2023.

IOT UK (2017) 'What's at the hear of a smart city? City Verve Manchester Roundtable', Digital Catapult, IoTUK.org.uk.

Istrate, R., Tulus, V., Grass, R. N., Vanbever, L., Stark, W. J., and Guillén-Gosálbez, G. (2024) 'The environmental sustainability of digital content consumption', *Nature Communications*, 15 : 3724.

Itten, R., Hischier, R., Andrae, A. S. G. et al (2020) 'Digital transformation – life cycle assessment of digital services, multifunctional devices and cloud computing', *The International Journal of Live Cycle Assessment*, 25: 2093–2098.

Jefferies, T. and Anderson, J. (2017) 'City Verve Internet of Things Demonstrator: City Verve Commissions: Manchester's Plinth', *Proceedings of Proceedings of EVA London.* https://doi.org/10.14236/ewic/EVA2017.55

Kaun, A. and Stiernstedt, F. (2020) 'Doing time, the smart way? Temporalities of the smart prison', *New Media & Society*, 22(9): 1580–1599.

Kim, K.-G. (2018) *Low-Carbon Smart Cities: Tools for Climate Resilience Planning.* Cham: Springer International Publishing.

Kitchin, R. (2014) 'The real-time city? Big data and smart urbanism', *Geojournal* 79(1): 1–14.

Kitchin, R. (2019) 'The timescape of smart cities', *Annals of the American Association of Geographers*, 109(3): 775–790.

Kitchin, R. (2023) *Digital Timescapes: Technology, Temporality and Society.* Cambridge: Polity.

Klöpffer, W. and Grahl, B. (2014) *Life Cycle Assessment (LCA): A Guide to Best Practice.* Whiley: Frankfurt am Main.

Knox, H. (2020) *Thinking Like a Climate Governing a City in Times of Environmental Change,* Durham: Duke University Press.

Kuntsman, A. (2007) '"Error: No such entry": Haunted ethnographies of on-line archives', *M/C: Journal of Media and Culture*, 10(5) October 2007, http://journal.media-culture.org.au/0710/, accessed 1 December 2023.

Kuntsman, A. (2020) 'Civil freedoms, collective memory and the environment in the age of digital inevitability: Rethinking digital futurities', in G. Asmolov (ed) *Information Technologies and the Civil Society*, Greenhouse of social technologies, Moscow (in Russian), https://hs.te-st.org/

Kuntsman, A. and Miyake, E. (2022) *Paradoxes of Digital Disengagement: In Search of the Opt Out Button.* London: Westminster University Press.

Kuntsman, A. and Rattle, I. (2019) 'Towards a paradigmatic shift in sustainability studies: A systematic review of peer reviewed literature and future agenda setting to consider environmental (un)sustainability of digital communication', *Environmental Communication*, 13(5): 567–581.

Liboiron, M. and Lepawsky, J. (2022) *Discard Studies: Wasting, Systems and Power*. London: MIT Press.

Liu, X. (2017) 'Air quality index as the stuff of the political', *Australian Feminist Studies*, 32(94): 445–460.

Liu, X. (2019) 'Nose hair: Love it or leave it?: The lovecidal of bodies that filter', *Parallax*, 25(1): 75–91.

Lock, I. and Seele, P. (2017) 'Theorizing stakeholders of sustainability in the digital age', *Sustainability Science*, 12(2): 235–245.

Manchester City Council (2021) 'Manchester digital strategy: Creating an inclusive, sustainable and resilient smart city', revised copy for continuing consultation, 29 January 2021, https://www.manches ter.gov.uk/download/downloads/id/27935/digital_strategy.pdf, accessed June 2021.

Manchester City Council (2022) 'Manchester Digital Strategy 2021–2026', https://www.manchester.gov.uk/digitalstrategy https://www.manchester.gov.uk/downloads/download/7436/manche ster_digital_strategy_2021_-_2026, accessed 15 March 2023.

Marks, L. (2020) 'Streaming video, a surprising link between pandemic and climate crisis', *Journal of Visual Culture*, Harun Farocki Institut special issue on Covid-19, April 2020, https://www.harun-farocki-institut.org", /en/2020/04/16/streaming-video-a-link-between-pandemic-and-climate-crisis-journal-of-visual-culture-hafi-2/, accessed 1 March 2023.

Marks, L. (2023) 'Small files for a small world', *Can we 'reduce, reuse, refuse'.. the digital? Digital Politics Summer School*, 5–6 June, 2023, https://digitalpoliticsmanmet.bloggi.co/can-we-reduce-reuse-ref use-the-digital-digital-politics-summer-school-5-6-june-2023, accessed 1 December 2023.

Marx, K. (1867) *Capital: A Critique of Political Economy*, vol. 1, https://www.marxists.org/archive/marx/works/1867-c1/

Masu Planning (2017) 'Master Plan of the Nihti Neighbourhood, Helsinki FI', https://www.masuplanning.com/project/nihti-mas ter-plan/, accessed 12 November 2023.

Matschoss, K., and E. Heiskanen (2017) 'Making it experimental in several ways: The work of intermediaries in raising the ambition level in local climate initiatives', *Journal of Cleaner Production*, 169: 85–93.

Mavropoulos, A., Tsakona, M., and Anthouli, A. (2015) 'Urban waste management and the mobile challenge', *Waste Management & Research*, 33(4): 381–387.

McLaren, D. and Agyeman, J. (2015) *Sharing Cities: A Case for Truly Smart and Sustainable Cities*. Urban and Industrial Environments. Cambridge, MA: MIT Press.

Méndez, G. (2008) *The Reinvention of the Saltpeter Industry*. Santiago: SQM Department of Communication.

Miconi, A. (2023) 'On digital fetishism: A critique of the big data paradigm', *Critical Sociology*, 0(0), https://doi.org/10.1177/08969205231202873, accessed 31 May 2024.

Miller, V. and Garcia, G. C. (2019) 'Digital ruins', *Cultural Geographies*, 26(4): 435–454.

Moran, J. (2018) 'Earthrise: The story behind our planet's most famous photo', *The Guardian*, 22 December 2018.

Morozov, E. (2013) *To Save Everything, Click Here: Technology, Solutionism and the Urge to Fix Problems That Don't Exist*. London: Allen Lane.

Mouton, M. and Burns, R. (2021) '(Digital) Neo-colonialism in the smart city', *Regional Studies*, 55(12): 1890–1901.

Muench, S., Stoermer, E., Jensen, K., Asikainen, T., Salvi, M., and Scapolo, F. (2022) *Towards a Green and Digital Future*. Luxembourg: Publications Office of the European Union.

Muiderman, K., Gupta, A., Vervoort, J., and Biermann, F. (2020) 'Four approaches to anticipatory climate governance: Different conceptions of the future and implications for the present', *WIREs Climate Change*, 11: 1–20, https://doi.org/10.1002/wcc.673, accessed 1 March 2024.

Mukherjee, J. (2018) *Sustainable Urbanisation in India: Challenges and Opportunities*. Singapore: Springer Singapore.

Murphy, M. (2006) *Sick Building Syndrome and the Problem of Uncertainty: Environmental Politics, Technoscience, and Women Workers*. Durham: Duke University Press.

Murphy, M. (2015) 'Unsettling care: Troubling transnational itineraries of care in feminist health practices', *Social Studies of Science*, 45(5): 717–737.

Myco (n.d.) 'Our aims and values', *MYCO: Manchester Mushroom Co-op*, https://www.mycomanchester.com/values, accessed 28 November 2023.

Nader Sayún, M. (2020) 'What is a smart city? Watch Michel Nader Sayún's videoblogs about Smart Kalasatama for a fresh point of view', *Smart Kalasatama,* 20 February 2020, https://fiksukalasat ama.fi/en/what-is-a-smart-city-watch-michel-naders-videobl ogs-about-smart-kalasatama-for-a-fresh-point-of-view/, accessed 10 November 2023.

Ngai, S. (2020) *Theory of the Gimmick: Aesthetic Judgement and Capitalist Form.* London: The Belknap Press of Harvard University Press.

Nixon, R. (2011) *Slow Violence and the Environmentalism of the Poor.* Cambridge, MA: Harvard University Press.

Noble, S. U. (2018) *Algorithms of Oppression: How Search Engines Reinforce Racism.* New York: New York University Press.

Noortje, M. (2015) *Material Participation: Technology, the Environment and Everyday Publics,* London: Palgrave Macmillan.

Oxford Road Corridor (n.d.a) 'Oxford Road Corridor', https://oxfordroadcorridor.com/, accessed 1 June 2022.

Oxford Road Corridor (n.d.b) 'Digital and smart cities', https://oxfordroadcorridor.com/knowledge/smart-cities/, accessed 1 June 2022.

Oxford Road Corridor (n.d.c) 'Manchester Science Park', https://oxfordroadcorridor.com/venues/manchester-science-park/, accessed 1 June 2022.

Palmisano, S. (2008) 'A smart planet: The next leadership agenda', *Ideas from IBM,* https://www.ibm.com/ibm/ideasfromibm/za/en/smarterplanet/20081106/sjp_speech.shtml, accessed 1 October 2023.

Pomponi, F., Saint, R., Arehart, J. H., Gharavi, N., and D'Amico, B. (2021) 'Decoupling density from tallness in analysing the life cycle greenhouse gas emissions of cities'., *nprj Urban Sustainability*, 1(33): 1–10.

Qui, J. L. (2016) *Goodbye ISlave: A Manifesto for Digital Abolition.* Geopolitics of Information. Urbana, Chicago: University of Illinois Press.

Royal Academy of Engineering (n.d.) 'Case Study 2: CityVerve Manchester: A platform of platforms for smart city data sharing', *Towards Trusted Data Sharing: Guidance and Case Studies*, https://reports.raeng.org.uk/datasharing/case-study-2-cityverve-manchester, accessed January 2022.

Salkkonen, P. (2013) 'From unknown problem to an acknowledged local environmental problem – A case of polluted soil in the City of Helsinki', *Local Environment*, 18(8): 888–903.

Santiago, M. (2013) 'Extracting histories: Mining, workers and environment', *New Environmental Histories of Latin America and the Caribbean*, 81–88.

Savolainen, L. (2023) 'Dirty, toxic, dumped: Waste as data metaphor', in A. Kuntsman and L. Xin (eds) *Digital Politics, Digital Histories, Digital Futures (Digital Activism and Society: Politics, Economy and Culture in Network Communication)*. Leeds: Emerald Publishing Limited, 89–104.

Schaal, S. and Lude, A. (2015) 'Using mobile devices in environmental education and education for sustainable development – Comparing theory and practice in a nation wide survey', *Sustainability*, 7(8): 10153–10170.

Sengupta, S. and Sengupta, U. (in progress) 'Can Smart City Supply Chains be Sustainable? Technology Enhanced Supply Chain Traceability in Urban Planning'.

Sengupta, U. and Sengupta, S. (2022a) 'Why government supported smart city initiatives fail: Examining community risk and benefit agreements as a missing link to accountability for equity-seeking groups', *Frontiers in Sociology*, 4, https://doi.org/10.3389/frsc.2022.960400, accessed 1 April 2024.

Sengupta, U. and Sengupta, S. (2022b) 'SDG-11 and smart cities: Contradictions and overlaps between social and environmental justice research agendas', *Frontiers in Sociology*, 7, https://doi.org/10.3389/fsoc.2022.995603, accessed 1 April 2024.

Shelton, T., Zook, M., and Wiig, A. (2015) 'The "actually existing smart city"'. *Cambridge Journal of Regions, Economy and Society*, 8(1): 13–25.

Siemens (2019) '€30 Million Triangulum sustainable cities project reaches successful completion', 25 September 2019, https://news.siemens.co.uk/news/30-million-triangulum-sustainable-cities-project-reaches-successful-completion, accessed January 2023.

Slatcher, A. (2016) *Smart City Manchester*, Manchester City Council Report, https://portal.ogc.org/files/?artifact_id=84030, accessed 1 March 2023.

Smart Kalasatama (n.d.a) Homepage, https://fiksukalasatama.fi/en/, accessed 10 November 2023.

Smart Kalasatama (n.d.b) 'Internet of Things (IoT) trials', https://fiksukalasatama.fi/en/smart-city/internet-of-things-iot-trials-in-smart-kalasatama/, accessed 10 November 2023.

Smart Kalasatama (n.d.c) 'Smart solutions', https://fiksukalasatama.fi/en/building-blocks/, accessed 10 July 2023.

Speare-Cole, R. (2021) 'Biomass is promoted as a carbon neutral fuel. But is burning wood a step in the wrong direction?' *The Guardian*, 5 October 2021, https://www.theguardian.com/environment/2021/oct/04/biomass-plants-us-south-carbon-neutral, accessed 27 September 2023.

Starosielski, N. and Walker, J. (eds) (2016) *Sustainable Media: Critical Approaches to Media and Environment*. New York: Routledge.

Stevenson, M. and Helmond, A. (2020) 'Legacy systems: Internet histories of the abandoned, discontinued and forgotten', *Internet Histories*, 4(1): 1–5.

Thales (2021) 'Environment and the IoT – 5 cases', 12 January 2021, https://www.thalesgroup.com/en/markets/digital-identity-and-security/iot/magazine/five-ways-iot-helping-environment, accessed 1 April 2024.

The University of Manchester (n.d.a) 'Triangulum', https://www.digitalfutures.manchester.ac.uk/what_we_do/societal-challenges/cities-and-environment/projects/triangulum/, accessed 15 February 2023

The University of Manchester (n.d.b) 'Triangulum: Demonstrating smart green growth in urban areas', Impact Report, https://research.manchester.ac.uk/en/impacts/triangulum-demonstrating-smart-green-growth-in-urban-areas, accessed 1 December 2022.

Tien Vu, H., Viet Do, H., Seo, H., and Liu, Y. (2020) 'Who leads the conversation on climate change?: A study of a global network of NGOs on Twitter', *Environmental Communication*, 14(4): 450–464.

Tomar, P. and Kaur, G. (2020) *Green and Smart Technologies for Smart Cities*. Abingdon, UK: Routledge.

Turner, J. M. (2022) *Charged: A History of Batteries and Lessons for a Clean Energy Future*. Seattle: University of Washington Press.

TWI (n.d.) 'What is a smart city? – Definition and examples', *TWI*, https://www.twi-global.com/technical-knowledge/faqs/what-is-a-smart-city#:~:text=There%20are%20a%20number%20of,the%20use%20of%20limited%20resources, accessed 2 October 2023.

Twitter/X (2018) City Verve status updates, https://twitter.com/cityverve/status/1024245191547514881, accessed 3 January 2024.

United Nations Department of Economic and Social Affairs (2022). *Goal 11 Make Cities and Human Settlements Inclusive, Safe, Resilient and Sustainable*, https://sdgs.un.org/goals/goal11, accessed 22 September 2022.

Vadén, T., Majava, A., Toivanen, T., Järvensivu, P., Hakala, E., and Eronen J. T. (2019) 'To continue to burn something? Technological, economic and political path dependencies in district heating in Helsinki, Finland', *Energy Research & Social Science*, 58: 101270.

van Dijck, J., Poell, T., and de Waal, M. (2018) *The Platform Society: Public Values in a Connective World*. Oxford: Oxford University Press.

Velkova, J. (2016) 'Data that warms: Waste heat, infrastructural convergence and the computation traffic commodity', *Big Data & Society*, 3(2): 1–10.

Venkatachary, S. K., Prasad, J., and Samikannu, R. (2017) 'Challenges, opportunities and profitability in virtual power plant business models in Sub Saharan Africa-Botswana', *International Journal of Energy Economics and Policy*, 7(4): 48–58.

Vierikko, K., Lähde, E., Nyberg, E., Korpilo, S., and Raymond, C. (2022) 'Shifting concepts of urban landscape in Helsinki: From primary forests to high tech nature-based solutions', in A. Rastandeh and M. Jarchow (eds) *Creating Resilient Landscapes in an Era of Climate Change: Global Case Studies and Real-World Solutions.* New York: Routledge, 179–193.

Virilio, P. (2000) *The Information Bomb.* New York: Verso.

Wajcman, J. (2014) *Acceleration of Life in Digital Capitalism.* Chicago, IL: University of Chicago Press.

Wang, X. (2020) *Blockchain Chicken Farm: And Other Stories of Tech in China's Countryside.* FSG Originals and logics.

Warde, P. (2018) *The Invention of Sustainability: Nature and Destiny, c.1500–1870.* Cambridge: Cambridge University Press.

Warde, P., Libby, R., and Sörlin, S. (2018) *The Environment: A History of the Idea.* Baltimore: Johns Hopkins University Press.

Weber, R. (2019) 'Anticipatory knowledge: How development consultants see the future', in M. Raco, and F. Savini (eds) *Planning and knowledge: How New Forms of Technocracy Are Shaping Contemporary Cities.* Bristol: Bristol University Press/Policy Press, 91–102.

White, J. M. (2016) 'Anticipatory logics of the smart city's global imaginary', *Urban Geography*, 37(4): 572–589.

Wiig, A. (2015) 'The empty rhetoric of the smart city: From digital inclusion to economic promotion in Philadelphia', *Urban Geography*, 37(4): 535–553.

Williams, P. (2018) 'Being matrixed: The (over)policing of gang suspects in London', *Stop Watch: Research and Action for Fair and Inclusive Policing*, https://www.stop-watch.org/our-work/gangs-matrix, accessed 1 April 2023.

YouTube (2016) 'CityVerve kick-off: Building a blueprint for smarter cities', https://www.youtube.com/watch?v=jgTfEGy0 Cco, accessed 1 March 2023.

YouTube (2018) 'Manchester: City of firsts', https://www.youtube.com/watch?v=CwhkrTilH6A&t=2s, accessed 15 March 2023.

Zakharova, I. and Jarke, J. (2024) 'Care-ful data studies: Or, what do we see, when we look at datafied societies through the lens of care?' *Information, Communication & Society*, 27(4): 651–664.

Zheng, S., Sheng, B., Ghafoor, A., Ashraf, A. A., and Qamri, G. M. (2023) 'Investigating the environmental externalities of digital financial inclusion and the COVID-19 pandemic: An environmental sustainability perspective', *Environmental Science and Pollution Research*, 30: 80758–80767.

# Index

References to endnotes show both the page number and the note number (133n2).